Student Lab Manual

for
Argument-Driven Inquiry
in
PHYSICS
VOLUME 1

MECHANICS LAB INVESTIGATIONS
for GRADES 9–12

Student Lab Manual
for
Argument-Driven Inquiry
in
PHYSICS
VOLUME 1

MECHANICS LAB INVESTIGATIONS
for GRADES 9–12

Victor Sampson, Todd L. Hutner, Daniel FitzPatrick,
Adam LaMee, and Jonathon Grooms

NSTApress
National Science Teachers Association
Arlington, Virginia

National Science Teachers Association

Claire Reinburg, Director
Rachel Ledbetter, Managing Editor
Deborah Siegel, Associate Editor
Amanda Van Beuren, Associate Editor
Donna Yudkin, Book Acquisitions Manager

Art and Design
Will Thomas Jr., Director

Printing and Production
Catherine Lorrain, Director

National Science Teachers Association
David L. Evans, Executive Director
David Beacom, Publisher

1840 Wilson Blvd., Arlington, VA 22201
www.nsta.org/store
For customer service inquiries, please call 800-277-5300.

Library of Congress Cataloging-in-Publication Data are available from the Library of Congress.
LCCN: 2017011353
ISBN: 978-1-68140-579-7
e-ISBN: 978-1-68140-580-3

SUSTAINABLE FORESTRY INITIATIVE   Certified Sourcing   www.sfiprogram.org   SFI-00756

CONTENTS

About the Authors xi

SECTION 1

Introduction and Lab Safety

Introduction 3

Safety in the Science Classroom, Laboratory, and Field Sites 5

SECTION 2

Motion and Interactions: Kinematics

INTRODUCTION LABS

Lab 1. Acceleration and Velocity: How Does the Direction of Acceleration Affect the Velocity of an Object?
    Lab Handout 14
    Checkout Questions 20

Lab 2. Acceleration and Gravity: What Is the Relationship Between the Mass of an Object and Its Acceleration During Free Fall?
    Lab Handout 23
    Checkout Questions 29

Lab 3. Projectile Motion: How Do Changes to the Launch Angle, the Initial Velocity, and the Mass of a Projectile Affect Its Hang Time?
    Lab Handout 33
    Checkout Questions 38

APPLICATION LAB

Lab 4. The Coriolis Effect: How Do the Direction and Rate of Rotation of a Spinning Surface Affect the Path of an Object Moving Across That Surface?
    Lab Handout 42
    Checkout Questions 48

SECTION 3
Forces and Motion: Dynamics

INTRODUCTION LABS

Lab 5. Force, Mass, and Acceleration: What Is the Mathematical Relationship Among the Net Force Exerted on an Object, the Object's Inertial Mass, and Its Acceleration?
Lab Handout.. 54
Checkout Questions.. 61

Lab 6. Forces on a Pulley: How Does the Mass of the Counterweight Affect the Acceleration of a Pulley System?
Lab Handout.. 64
Checkout Questions.. 70

Lab 7. Forces on an Incline: What Is the Mathematical Relationship Between the Angle of Incline and the Acceleration of an Object Down the Incline?
Lab Handout.. 72
Checkout Questions.. 77

APPLICATION LABS

Lab 8. Friction: Why Are Some Lubricants Better Than Others at Reducing the Coefficient of Friction Between Metal Plates?
Lab Handout.. 82
Checkout Questions.. 89

Lab 9. Falling Objects and Air Resistance: How Does the Surface Area of a Parachute Affect the Force Due to Air Resistance as an Object Falls Toward the Ground?
Lab Handout.. 92
Checkout Questions.. 98

SECTION 4

Forces and Motion: Circular Motion and Rotation

INTRODUCTION LABS

Lab 10. Rotational Motion: How Do the Mass and the Distribution of Mass in an Object Affect Its Rotation?
   Lab Handout.. 104
   Checkout Questions.. 110

Lab 11. Circular Motion: How Does Changing the Angular Velocity of the Swinging Mass at the Top of a Whirligig and the Amount of Mass at the Bottom of a Whirligig Affect the Distance From the Top of the Tube to the Swinging Mass?
   Lab Handout.. 113
   Checkout Questions.. 120

APPLICATION LAB

Lab 12. Torque and Rotation: How Can Someone Predict the Amount of Force Needed to Open a Bottle Cap?
   Lab Handout.. 124
   Checkout Questions.. 129

SECTION 5

Forces and Motion: Oscillations

INTRODUCTION LABS

Lab 13. Simple Harmonic Motion and Pendulums: What Variables Affect the Period of a Pendulum?
   Lab Handout.. 136
   Checkout Questions.. 141

Lab 14. Simple Harmonic Motion and Springs: What Is the Mathematical Model of the Simple Harmonic Motion of a Mass Hanging From a Spring?
   Lab Handout.. 144
   Checkout Questions.. 150

APPLICATION LAB

Lab 15. Simple Harmonic Motion and Rubber Bands: Under What Conditions Do Rubber Bands Obey Hooke's Law?
Lab Handout.. 156
Checkout Questions.. 161

SECTION 6

Forces and Motion: Systems of Particles and Linear Momentum

INTRODUCTION LABS

Lab 16. Linear Momentum and Collisions: When Two Objects Collide and Stick Together, How Do the Initial Velocity and Mass of One of the Moving Objects Affect the Velocity of the Two Objects After the Collision?
Lab Handout.. 168
Checkout Questions.. 175

Lab 17. Impulse and Momentum: How Does Changing the Magnitude and Duration of a Force Acting on an Object Affect the Momentum of That Object?
Lab Handout.. 179
Checkout Questions.. 185

APPLICATION LABS

Lab 18. Elastic and Inelastic Collisions: Which Properties of a System Are Conserved During a Collision?
Lab Handout.. 190
Checkout Questions.. 196

Lab 19. Impulse and Materials: Which Material Is Most Likely to Provide the Best Protection for a Phone That Has Been Dropped?
Lab Handout.. 200
Checkout Questions.. 206

SECTION 7

Energy, Work, and Power

INTRODUCTION LABS

Lab 20. Kinetic and Potential Energy: How Can We Use the Work-Energy Theorem to Explain and Predict Behavior of a System That Consists of a Ball, a Ramp, and a Cup?
Lab Handout... 212
Checkout Questions... 218

Lab 21. Conservation of Energy and Pendulums: How Does Placing a Nail in the Path of a Pendulum Affect the Height of a Pendulum Swing?
Lab Handout... 222
Checkout Questions... 227

Lab 22. Conservation of Energy and Wind Turbines: How Can We Maximize the Amount of Electrical Energy That Will Be Generated by a Wind Turbine Based on the Design of Its Blades?
Lab Handout... 230
Checkout Questions... 236

APPLICATION LAB

Lab 23. Power: Which Toy Car Has the Engine With the Greatest Horsepower?
Lab Handout 240
Checkout Questions... 245

Image Credits 249

ABOUT THE AUTHORS

Victor Sampson is an associate professor of STEM (science, technology, engineering, and mathematics) education and the director of the Center for STEM Education (see *http://stemcenter.utexas.edu*) at The University of Texas at Austin (UT-Austin). He received a BA in zoology from the University of Washington, an MIT from Seattle University, and a PhD in curriculum and instruction with a specialization in science education from Arizona State University. Victor also taught high school biology and chemistry for nine years. He specializes in argumentation in science education, teacher learning, and assessment. To learn more about his work in science education, go to *www.vicsampson.com*.

Todd L. Hutner is the assistant director for teacher education and center development for the Center of STEM Education at UT-Austin. He received a BS and an MS in science education from Florida State University (FSU) and a PhD in curriculum and instruction from UT-Austin. Todd's classroom teaching experience includes teaching chemistry, physics, and Advanced Placement (AP) physics in Texas and earth science and astronomy in Florida. His current research focuses on the impact of both teacher education and education policy on the teaching practice of secondary science teachers.

Daniel FitzPatrick is a clinical assistant professor and master teacher in the UTeach program at the UT-Austin. He received a BS and an MA in mathematics from UT Austin and is currently a doctoral student in STEM education. Prior to his work in higher education, Daniel taught both middle and high school mathematics in public and charter schools. His interests include argumentation in mathematics education, teacher preparation, and the use of dynamic software in teaching and learning mathematics.

Adam LaMee has taught high school physics in Florida for 12 years and is currently the PhysTEC teacher-in-residence at the University of Central Florida (*http://sciences. ucf.edu/physics/phystec*). He received a BS in physics and a BS in anthropology from FSU. Adam is a Quarknet Teaching and Learning Fellow. He also contributed to the development of Florida's state science education standards and teacher certification exams, worked on the CERN's CMS (Compact Muon Solenoid) experiment, and has researched game-based assessment and performance assessment alternatives to large-scale testing. Learn more about his work at *www.adamlamee.com*.

Jonathon Grooms is an assistant professor of curriculum and pedagogy in the Graduate School of Education and Human Development at The George Washington University. He received a BS in secondary science and mathematics teaching with a focus in chemistry and physics from FSU. Upon graduation, Jonathon joined FSU's

Office of Science Teaching, where he directed the physical science outreach program Science on the Move. He also earned a PhD in science education from FSU. To learn more about his work in science education, go to *www.jgrooms.com*.

SECTION 1
Introduction and Lab Safety

INTRODUCTION

Science is much more than a collection of facts or ideas that describe what we know about how the world works and why it works that way. Science is also a set of crosscutting concepts and practices that people can use to develop and refine new explanations for, or descriptions of, the natural world. These core ideas, crosscutting concepts, and scientific practices are important for you to learn. When you understand these core ideas, crosscutting concepts, and scientific practices, it is easier to appreciate the beauty and wonder of science, to engage in public discussions about science, and to critique the merits of scientific findings that are presented through the popular media. You will also have the knowledge and skills that you will need to continue to learn about science outside school or to enter a career in science, engineering, or technology once you learn the core ideas, crosscutting concepts, and scientific practices.

The core ideas of science include the theories, laws, and models that scientists use to explain natural phenomena or to predict the results of new investigations. The crosscutting concepts of science, in contrast, are themes that have value in every discipline of science as a way to help understand a natural phenomenon. These crosscutting concepts can be used as an organizational framework for connecting knowledge from the various fields of science into a coherent and scientifically based view of the world and to help us think about what is important to think about or look for during an investigation. Finally, scientific practices are used to develop and refine new ideas about the world. Although some practices differ from one field of science to another, all fields share a set of common practices. The practices including such things as asking and answering questions, planning and carrying out investigations, analyzing and interpreting data, and obtaining, evaluating, and communicating information. One of the most important scientific practices is arguing from evidence. Arguing from evidence, or the process of proposing, supporting, challenging, and refining claims based on evidence, is important because scientists need to be able to examine, review, and evaluate their own knowledge and ideas and critique those of others. Scientists also argue from evidence when they need to appraise the quality of data, develop and refine models, develop new testable questions from those models, and suggest ways to refine or modify existing theories, laws, and models.

It is important to always remember that science is a social activity, not an individual one. Science is a social activity because many different scientists contribute to the development of new scientific knowledge. As scientists carry out their research, they frequently talk with their colleagues, both formally and informally. They exchange e-mails, engage in discussions at conferences, share research techniques and analytical procedures, and present new ideas by writing articles in journals or chapters in books. They also critique the ideas and methods used by other scientists through a formal peer-review process before they can be published in journals or books. In short, scientists are members of a community, and the members of that community work together to build, develop, test, critique, and refine ideas. The ways scientists talk, write, think, and interact with each other reflect common ideas about what counts as quality and shared standards for how new ideas should be developed, shared, evaluated, and refined. These ways of talking, writing, thinking, and interacting make science different from other ways of knowing. The core ideas, crosscutting concepts, and scientific practices are important within the scientific community because most, if not all, of the members of that community find them to be a useful way to develop and refine new explanations for, or descriptions of, the natural world.

The laboratory investigations that are included in this book are designed to help you learn the core ideas, crosscutting concepts, and scientific practices that are important in physics. During each investigation, you will have an opportunity to use a core idea and several crosscutting concepts and scientific practices in order to understand a natural phenomenon or to solve a problem. Your teacher will introduce each investigation by giving you a task to accomplish and a guiding question to answer. You will then work as part of a team to plan and carry out an investigation to collect the data that you need to answer that question. From there, your team will develop an initial argument that includes a claim, evidence in support of your claim, and a justification of your evidence. The claim will be your answer to the guiding question, the evidence will include your analysis of the data that you collected and an interpretation of your analysis, and the justification will explain why your evidence is important. Next, you will have an opportunity to share your argument with your classmates and critique their arguments, much like professional scientists do. You will then revise your initial argument based on their feedback. Finally, you will be asked to write an investigation report on your own to share what you learned. These reports will go through a double-blind peer review so you can improve it before you submit to your teacher for a grade. As you complete more and more investigations in this lab manual, you will not only learn the core ideas associated with each investigation but also get better at using crosscutting concepts and scientific practices to understand the natural world.

SAFETY IN THE SCIENCE CLASSROOM, LABORATORY, AND FIELD SITES

Note to science teachers and supervisors/administrators: The following safety acknowledgment form is for your use in the classroom and should be given to students at the beginning of the school year to help them understand their role in ensuring a safer and productive science experience.

Science is a process of discovering and exploring the natural world. Exploration occurs in the classroom/laboratory or in the field. As part of your science class, you will be doing many activities and investigations that will involve the use of various materials, equipment, and chemicals. Safety in the science classroom, laboratory, or field sites is the FIRST PRIORITY for students, instructors, and parents. To ensure safer classroom/laboratory/field experiences, the following **Science Rules and Regulations** have been developed for the protection and safety of all. Your instructor will provide additional rules for specific situations or settings. The rules and regulations must be followed at all times. After you have reviewed them with your instructor, read and review the rules and regulations with your parent/guardian. Their signature and your signature on the safety acknowledgment form are required before you will be permitted to participate in any activities or investigations. Your signature indicates that you have read these rules and regulations, understand them, and agree to follow them at all times while working in the classroom/laboratory or in the field.

Safety Standards of Student Conduct in the Classroom, Laboratory, and in the Field

1. Conduct yourself in a responsible manner at all times. Frivolous activities, mischievous behavior, throwing items, and conducting pranks are prohibited.

2. Lab and safety information and procedures must be read ahead of time. All verbal and written instructions shall be followed in carrying out the activity or investigation.

3. Eating, drinking, gum chewing, applying cosmetics, manipulating contact lenses, and other unsafe activities are not permitted in the laboratory.

4. Working in the laboratory without the instructor present is prohibited.

5. Unauthorized activities or investigations are prohibited. Unsupervised work is not permitted.

6. Entering preparation or chemical storage areas is prohibited at all times.

7. Removing chemicals or equipment from the classroom or laboratory is prohibited unless authorized by the instructor.

Personal Safety

8. Sanitized indirectly vented chemical splash goggles or safety glasses as appropriate (meeting the ANSI Z87.1 standard) shall be worn during activities or demonstrations in the classroom, laboratory, or field, including pre-laboratory work and clean-up, unless the instructor specifically states that the activity or demonstration does not require the use of eye protection.

9. When an activity requires the use of laboratory aprons, the apron shall be appropriate to the size of the student and the hazard associated with the activity or investigation. The apron shall remain tied throughout the activity or investigation.

10. All accidents, chemical spills, and injuries must be reported immediately to the instructor, no matter how trivial they may seem at the time. Follow your instructor's directions for immediate treatment.

11. Dress appropriately for laboratory work by protecting your body with clothing and shoes. This means that you should use hair ties to tie back long hair and tuck into the collar. Do not wear loose or baggy clothing or dangling jewelry on laboratory days. Acrylic nails are also a safety hazard near heat sources and should not be used. Sandals or open-toed shoes are not to be worn during any lab activities. Refer to pre-lab instructions. If in doubt, ask!

12. Know the location of all safety equipment in the room. This includes eye wash stations, the deluge shower, fire extinguishers, the fume hood, and the safety blanket. Know the location of emergency master electric and gas shut offs and exits.

13. Certain classrooms may have living organisms including plants in aquaria or other containers. Students must not handle organisms without specific instructor authorization. Wash your hands with soap and water after handling organisms and plants.

14. When an activity or investigation requires the use of laboratory gloves for hand protection, the gloves shall be appropriate for the hazard and worn throughout the activity.

Specific Safety Precautions Involving Chemicals and Lab Equipment

15. Avoid inhaling fumes that may be generated during an activity or investigation.

16. Never fill pipettes by mouth suction. Always use the suction bulbs or pumps.

17. Do not force glass tubing into rubber stoppers. Use glycerin as a lubricant and hold the tubing with a towel as you ease the glass into the stopper.

18. Proper procedures shall be followed when using any heating or flame-producing device, especially gas burners. Never leave a flame unattended.

19. Remember that hot glass looks the same as cold glass. After heating, glass remains hot for a very long time. Determine if an object is hot by placing your hand close to the object but do not touch it.

20. Should a fire drill, lockdown, or other emergency occur during an investigation or activity, make sure you turn off all gas burners and electrical equipment. During an evacuation emergency, exit the room as directed. During a lockdown, move out of the line of sight from doors and windows if possible or as directed.

21. Always read the reagent bottle labels twice before you use the reagent. Be certain the chemical you use is the correct one.

22. Replace the top on any reagent bottle as soon as you have finished using it and return the reagent to the designated location.

23. Do not return unused chemicals to the reagent container. Follow the instructor's directions for the storage or disposal of these materials.

Standards for Maintaining a Safer Laboratory Environment

24. Backpacks and books are to remain in an area designated by the instructor and shall not be brought into the laboratory area.

25. Never sit on laboratory tables.

26. Work areas should be kept clean and neat at all times. Work surfaces are to be cleaned at the end of each laboratory or activity.

27. Solid chemicals, metals, matches, filter papers, broken glass, and other materials designated by the instructor are to be deposited in the proper waste containers, not in the sink. Follow your instructor's directions for disposal of waste.

28. Sinks are to be used for the disposal of water and those solutions designated by the instructor. Other solutions must be placed in the designated waste disposal containers.

29. Glassware is to be washed with hot, soapy water and scrubbed with the appropriate type and sized brush, rinsed, dried, and returned to its original location.

30. Goggles are to be worn during the activity or investigation, clean up, and through hand washing.

31. Safety Data Sheets (SDSs) contain critical information about hazardous chemicals of which students need to be aware. Your instructor will review the salient points on the SDSs for the hazardous chemicals students will be working with and also post the SDSs in the lab for future reference.

Safety Acknowledgment Form: Science Rules and Regulations

I have read the science rules and regulations in the *Student Lab Manual for Argument-Driven Inquiry in Physics, Volume 1,* and I agree to follow them during any science course, investigation, or activity. By signing this form, I acknowledge that the science classroom, laboratory, or field sites can be an unsafe place to work and learn. The safety rules and regulations are developed to help prevent accidents and to ensure my own safety and the safety of my fellow students. I will follow any additional instructions given by my instructor. I understand that I may ask my instructor at any time about the rules and regulations if they are not clear to me. My failure to follow these science laboratory rules and regulations may result in disciplinary action.

_____   _____

Student Signature   Date

_____   _____

Parent/Guardian Signature   Date

SECTION 2
Motion and Interactions

Kinematics

Introduction Labs

Lab Handout

Lab 1. Acceleration and Velocity: How Does the Direction of Acceleration Affect the Velocity of an Object?

Introduction

The ability to describe motion is the basis of much of physics. To explain why objects move the way they do, we must first be able to describe how an object moves. *Velocity* and *acceleration* are terms often used to describe motion. These two terms, however, have different meanings in science. The rate of change in an object's position is called its velocity. An object's velocity describes how fast it travels and its direction of motion. The rate of change of velocity is called acceleration. Like position, velocity and acceleration are vector quantities with a magnitude (i.e., how much) and a direction (i.e., which way). Choosing a frame of reference when you study an object's motion is up to you. Your reference frame will specify which direction is the positive direction and which one is negative.

Graphs that show a change in position of an object over time (called a position vs. time graph) and the change in velocity of an object over time (called a velocity vs. time graph) can help us describe the motion of an object. On these types of graphs, the rate of change for the object of interest corresponds to the slope of the line. For example, the slope of the line on a position vs. time graph at a specific time is the velocity of an object at that time, because velocity is the rate of change of an object's position with respect to time. For a graph with a curved line, such as those shown in Figures L1.1 and L1.2, the slope of the line at a given point in time on the graph is defined as the slope of the tangent line to the curve at that specific point.

Not all rates of change remain constant, however. For example, it is possible for an object to increase or decrease in velocity. We can also represent this change on a motion graph. If the slope of the tangent line gets steeper or gets less steep downward as time increases, this is called a "concave up" graph (see Figure L1.1). A concave up position versus time graph indicates that an object's velocity is increasing over time. When

A concave up position vs. time graph indicates that the velocity of an object is increasing.

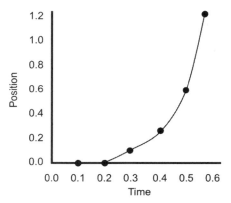

A concave down position vs. time graph indicates that the velocity of an object is decreasing.

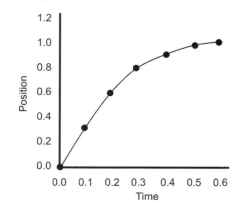

the slopes of the tangent lines decrease, in contrast, it is called a "concave down" graph. A concave down position versus time graph can either be getting less steep upward or becoming more steep downward (see Figure L1.2). Either way, a concave down position versus time graph indicates that the velocity of the object is decreasing over time.

Physicists use the term acceleration to describe any change in the velocity of an object. The acceleration of an object can be either positive or negative. The direction of that acceleration, however, is important because it may cause an object's velocity to change differently. In this investigation you will have an opportunity to explore the motion of a cart on a track in order to explain how the movement of that cart changes when it accelerates in different directions.

Your Task

You will observe the motion of a cart and then use what you know about vectors and graphs, patterns, and scale, proportion, and quantity to determine a mathematical relationship between velocity and acceleration.

The guiding question of this investigation is, *How does the direction of acceleration affect the velocity of an object?*

Materials

You may use any of the following materials during your investigation (some items may not be available):

- Safety glasses or goggles (required)
- Cart with fan attachment
- Track for cart
- Video camera
- Motion detector/sensor and interface
- Computer or tablet with data collection and analysis software and/or video analysis software
- Meterstick

Safety Precautions

Follow all normal lab safety rules. In addition, take the following safety precautions:

1. Wear sanitized safety glasses or goggles during lab setup, hands-on activity, and takedown.

2. Keep fingers and toes out of the way of moving objects.

3. Do not place fingers into the fan.

4. Keep hair, clothing, and jewelry away from the cart while the fan attachment is switched on.

5. Wash hands with soap and water after completing the lab.

LAB 1

Getting Started

To answer the guiding question, you will need to design and carry out an experiment using a fan cart and a track. The cart can be placed on the track so it will only move in two directions (i.e., left or right). You can use the fan to apply a force to the cart so it will accelerate in different directions. You have two options for tracking the motion of the cart on the track. The first option is to use a motion detector/sensor attached to the end of the track. The motion detector/sensor will allow you to record the exact position of the cart on the track at a given point in time and how the position of the cart relative to the motion detector/sensor changes on the track over time. Figure L1.3 shows how you can set up the equipment for this investigation if you use a motion detector/sensor to track the motion of the cart. The second option for tracking the motion of a cart is to use video analysis software. In this case, you will need to use a video camera and a meterstick. Place the meterstick in front of the track and then video record the cart as it moves. You can then upload the video to a computer or tablet and use video analysis software to examine the motion of the cart. The meterstick is important because it will provide a reference point in the video so you can measure the cart's displacement over time. Figure L1.4 shows how you can set up the equipment for this investigation if you decide to use video analysis software to track the motion of the cart.

FIGURE L1.3 _____

How to examine the motion of a cart using a motion sensor

FIGURE L1.4 _____

How to examine the motion of a cart using a video camera

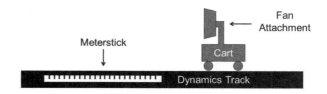

Before you can begin to design an experiment using this equipment, however, you will need to determine what type of data you need to collect, how you will collect it, and how you will analyze it. To determine what type of data you need to collect, think about the following questions:

- What are the boundaries and components of the system you are studying?
- Which factor(s) might control the rate of change in this system?
- How could you keep track of changes in this system quantitatively?

- What information do you need to determine the cart's velocity and acceleration?
- What will be the reference frame?
- Are your variables vector quantities or scalar quantities?

To determine *how you will collect the data*, think about the following questions:

- How will you change the direction of acceleration?
- What comparisons will you need to make?
- What measurement scale or scales should you use to collect data?
- How will you make sure that your data are of high quality (i.e., how will you reduce error)?
- For any vector quantities, which directions are positive and which directions are negative?

To determine *how you will analyze the data,* think about the following questions:

- How will you identify both the magnitude and direction of the vector quantities?
- What type of calculations will you need to make?
- How could you use mathematics to describe a change over time?
- How could you use mathematics to describe a relationship between variables?
- What types of patterns could you look for in your data?

Connections to the Nature of Scientific Knowledge and Scientific Inquiry

As you work through your investigation, you may want to consider

- the difference between observations and inferences in science, and
- the nature and role of experiments in science.

Initial Argument

Once your group has finished collecting and analyzing your data, your group will need to develop an initial argument. Your initial argument needs to include a claim, evidence supporting your claim, and a justification of the evidence. The claim is your group's answer to the guiding question. The evidence is an analysis and interpretation of your data. Finally, the justification of the evidence is why your group thinks the evidence matters. The justification of the evidence is important because scientists can use different kinds of evidence to support their claims. Your group will create your initial argument on a whiteboard. Your whiteboard should include all the information shown in Figure L1.5 (p. 18).

LAB 1

Argumentation Session

The argumentation session allows all of the groups to share their arguments. One or two members of each group will stay at the lab station to share that group's argument, while the other members of the group go to the other lab stations to listen to and critique the other arguments. This is similar to what scientists do when they propose, support, evaluate, and refine new ideas during a poster session at a conference. If you are presenting your group's argument, your goal is to share your ideas and answer questions. You should also keep a record of the critiques and suggestions made by your classmates so you can use this feedback to make your initial argument stronger. You can keep track of specific critiques and suggestions for improvement that your classmates mention in the space below.

Critiques about our initial argument and suggestions for improvement:

Argument presentation on a whiteboard

The Guiding Question:	
Our Claim:	
Our Evidence:	Our Justification of the Evidence:

If you are critiquing your classmates' arguments, your goal is to look for mistakes in their arguments and offer suggestions for improvement so these mistakes can be fixed. You should look for ways to make your initial argument stronger by looking for things that the other groups did well. You can keep track of interesting ideas that you see and hear during the argumentation in the space below. You can also use this space to keep track of any questions that you will need to discuss with your team.

Interesting ideas from other groups or questions to take back to my group:

Once the argumentation session is complete, you will have a chance to meet with your group and revise your initial argument. Your group might need to gather more data or design a way to test one or more alternative claims as part of this process. Remember, your goal at this stage of the investigation is to develop the best argument possible.

Report

Once you have completed your research, you will need to prepare an investigation report that consists of three sections. Each section should provide an answer to the following questions:

1. What question were you trying to answer and why?

2. What did you do to answer your question and why?

3. What is your argument?

Your report should answer these questions in two pages or less. This report must be typed, and any diagrams, figures, or tables should be embedded into the document. Be sure to write in a persuasive style; you are trying to convince others that your claim is acceptable or valid!

LAB 1

Lab 1. Acceleration and Velocity: How Does the Direction of Acceleration Affect the Velocity of an Object?

1. Two carts undergo positive acceleration, but their velocities are in opposite directions.

    a. Sketch each cart and label each one with arrows representing the directions of its velocity and its acceleration

    b. Sketch a single velocity versus time graph representing each cart.

    c. Sketch a single position versus time graph representing each cart.

2. Positive acceleration will cause an object to speed up.

   a. I agree with this statement.
   b. I disagree with this statement.

   Explain your answer, using an example from your investigation about the direction of acceleration and velocity.

3. *Observations* and *inferences* are terms that have the same meaning in science.

   a. I agree with this statement.
   b. I disagree with this statement.

   Explain your answer, using an example from your investigation about the direction of acceleration and velocity.

4. Scientists always design and carry out an experiment to answer scientific questions.

   a. I agree with this statement.
   b. I disagree with this statement.

   Explain your answer, using an example from your investigation about the direction of acceleration and velocity.

5. Why is it useful to identify patterns during an investigation? In your answer, be sure to include examples from at least two different investigations.

6. How are vector quantities and scalar quantities different in science? In your answer, be sure to include examples from at least two different investigations.

Lab Handout

Lab 2. Acceleration and Gravity: What Is the Relationship Between the Mass of an Object and Its Acceleration During Free Fall?

Introduction

The motion of an object is the result of all the different forces that are acting on the object. If you push a toy car across the floor, it moves in the direction you pushed it. If the car then hits a wall, the force of the wall causes the car to stop. Applying a push or a pull to an object is an example of a contact force, where one object applies a force to another object through direct contact. There are other types of forces that can act on objects that do not involve objects touching. For example, a strong magnet can pull on a paper clip and make it move without ever actually touching the paper clip. Another example is static electricity. Static electricity in a rubber balloon can cause a person's hair to stand up without the balloon actually touching any of his or her hair. Magnetic forces and electrical forces are therefore called non-contact forces. Perhaps the most common non-contact force is gravity. Gravity is a force of attraction between two objects; the force due to gravity always works to bring objects closer together.

Any two objects that have mass (and, remember, all matter has mass) will also experience a gravitational force of attraction between them. Consider the Sun, the Earth, and the Moon as examples. The Earth and the Moon are very large and have a lot of mass; the force of gravity between the Earth and the Moon is strong enough to keep the Moon orbiting the Earth even though they are very far apart. Similarly, the force of gravity between the Sun and the Earth is strong enough to keep the Earth in orbit around the Sun, despite the Earth and the Sun being millions of miles apart. The force of gravity between two objects depends on the amount of mass of each object and how far apart they are. Objects that are more massive produce a greater gravitational force. The force of gravity between two objects also weakens as the distance between the two objects increases. So even though the Earth and the Sun are very far apart from each other (which means less gravity), the fact that they are both very massive (which means more gravity) results in a gravitational force that is strong enough to keep the Earth in orbit.

The gravitational force that acts between any two objects, as noted earlier, can cause one of those objects to move. For example, a cell phone in free fall (see Figure L2.1) moves toward the center of the Earth because of gravity. Scientists describe the motion of an object in free fall by describing its velocity and acceleration. Velocity is the speed (distance in a specific amount of time) of an object in a given direction. Acceleration is the rate of change in velocity per unit time, most often the rate of change in velocity per second. The amount of force required to produce a specific acceleration in the motion of an object, such as a cell phone, depends on the mass of that object. Therefore, as the mass of an object increases, so does the amount of force that is needed to produce a specific acceleration.

In this investigation you will have an opportunity to explore the relationship between the mass of an object and its acceleration due to gravity during free fall. Many people think that heavier objects accelerate toward the ground faster than lighter ones because gravity will act on heavier objects with more force. Heavier objects will therefore have a greater acceleration because of the force of gravity. Other people, however, think that heavier objects have more inertia (the tendency of an object to resist changes in its motion) so heavier objects will be less responsive to the force of gravity and have a smaller acceleration. Still others think that mass of a falling object has no effect on acceleration due to gravity because the magnitude of the force of gravity acting on a falling object is dependent on the mass of that object. In this case, the greater force of gravity for a more massive object is countered by the greater inertia of the massive object, thereby resulting in an acceleration due to gravity that is unaffected by the falling object's mass.

A cell phone in free fall

Unfortunately, it is challenging to determine which of these three explanations is the most valid because objects encounter air resistance as they fall. Air resistance is the result of an object moving through a layer of air and colliding with air molecules. The more air molecules that an object collides with, the greater the air resistance force. Air resistance is therefore dependent on the velocity of the falling object and the cross-sectional surface area of the falling object. Since heavier objects are often larger than lighter ones (consider a bowling ball and a marble as an example), it is often difficult to design a fair test of these three explanations. To determine the relationship between mass and acceleration due to gravity, you will therefore need to design an experiment that will allow you to control for the influence of air resistance.

Your Task

Use what you know about forces and motion, patterns, and rates of change to design and carry out an experiment to determine the relationship between mass and acceleration due to gravity.

The guiding question of this investigation is, *What is the relationship between the mass of an object and its acceleration during free fall?*

Materials

You may use any of the following materials during your investigation:

- Safety glasses or goggles (required)
- Beanbags A, B, C, D, and/or E
- Meterstick
- Stopwatch
- Electronic or triple beam balance
- Masking tape

If you have access to the following equipment, you may also consider using a video camera and a computer or tablet with video analysis software.

Safety Precautions

Follow all normal lab safety rules. In addition, take the following safety precautions:

1. Wear sanitized safety glasses or goggles during lab setup, hands-on activity, and takedown.

2. Do not throw the beanbags.

3. Do not stand on tables or chairs.

4. Wash hands with soap and water after completing the lab.

Investigation Proposal Required? ☐ Yes   ☐ No

Getting Started

To answer the guiding question, you will need to design and conduct an experiment as part of your investigation. To accomplish this task, you must determine what type of data you need to collect, how you will collect it, and how you will analyze it.

To determine *what type of data you need to collect,* think about the following questions:

- What information will you need to be able to determine the acceleration of a falling object?

- Which factor(s) might control the rate of change in the velocity (i.e., acceleration) of a falling object?

- What will be the independent variable and the dependent variable for your experiment?

- Will you measure acceleration directly or will you have to calculate it using other measurements?

To determine *how you will collect the data,* think about the following questions:

- What variables will need to be controlled and how will you control them?

- How many tests will you need to run to have reliable data (to make sure it is consistent)?
- How will you make sure that your data are of high quality (i.e., how will you reduce error)?
- How will you keep track of the data you collect and how will you organize it?

To determine *how you will analyze the data,* think about the following questions:

- How will you calculate the acceleration of a falling object?
- What types of patterns might you look for as you analyze the data you collected?
- What type of calculations will you need to make to take into account multiple trials?
- What types of graphs or tables could you create to help make sense of your data?

Connections to the Nature of Scientific Knowledge and Scientific Inquiry

As you work through your investigation, you may want to consider

- the difference between laws and theories in science, and
- the difference between data and evidence in science.

Initial Argument

Once your group has finished collecting and analyzing your data, your group will need to develop an initial argument. Your initial argument needs to include a claim, evidence to support your claim, and a justification of the evidence. The claim is your group's answer to the guiding question. The evidence is an analysis and interpretation of your data. Finally, the justification of the evidence is why your group thinks the evidence matters. The justification of the evidence is important because scientists can use different kinds of evidence to support their claims. Your group will create your initial argument on a whiteboard. Your whiteboard should include all the information shown in Figure L2.2.

FIGURE L2.2

Argument presentation on a whiteboard

The Guiding Question:	
Our Claim:	
Our Evidence:	Our Justification of the Evidence:

Argumentation Session

The argumentation session allows all of the groups to share their arguments. One or two members of each group will stay at the lab station to share that group's argument, while the other members of the group go to the other lab stations to listen to and critique the other arguments. This is similar to what scientists do when they propose, support, evaluate, and refine new ideas during a poster session at a conference. If you are presenting your group's argument, your goal is to share your ideas and answer questions. You should also keep a record of the critiques and suggestions made by your classmates so you can use this feedback to make your initial argument stronger. You can keep track of specific critiques and suggestions for improvement that your classmates mention in the space below.

Critiques about our initial argument and suggestions for improvement:

If you are critiquing your classmates' arguments, your goal is to look for mistakes in their arguments and offer suggestions for improvement so these mistakes can be fixed. You should look for ways to make your initial argument stronger by looking for things that the other groups did well. You can keep track of interesting ideas that you see and hear during the argumentation in the space below. You can also use this space to keep track of any questions that you will need to discuss with your team.

Interesting ideas from other groups or questions to take back to my group:

LAB 2

Once the argumentation session is complete, you will have a chance to meet with your group and revise your initial argument. Your group might need to gather more data or design a way to test one or more alternative claims as part of this process. Remember, your goal at this stage of the investigation is to develop the best argument possible.

Report

Once you have completed your research, you will need to prepare an investigation report that consists of three sections. Each section should provide an answer to the following questions:

1. What question were you trying to answer and why?

2. What did you do to answer your question and why?

3. What is your argument?

Your report should answer these questions in two pages or less. This report must be typed, and any diagrams, figures, or tables should be embedded into the document. Be sure to write in a persuasive style; you are trying to convince others that your claim is acceptable or valid!

National Science Teachers Association

Checkout Questions

Lab 2. Acceleration and Gravity: What Is the Relationship Between the Mass of an Object and Its Acceleration During Free Fall?

For questions 1–3, assume air resistance can be neglected.

1. The picture at right shows a tall building. A physics class is investigating the time it takes for different objects that are dropped out of a window to fall to the ground. They release four objects from rest. Object A (mass = 5 kg) and object B (mass = 10 kg) are dropped from level 8. Object C (mass = 5 kg) and object D (mass = 10 kg) are dropped from level 4. What is the order in which they hit the ground?

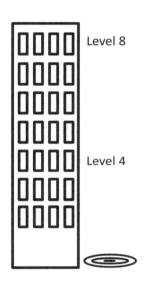

    a. Object D, then object C, then object B, then object A

    b. Object D, then objects C and B at the same time, then object A.

    c. Object D, then object B, then object C, then object A

    d. Objects C and D at the same time, then objects A and B at the same time

    How do you know?

2. Determine the amount of time that object A would take to fall if level 8 is 50 meters above the ground.

3. How long would it take object C to hit the ground if level 4 is half as high as level 8?

4. A theory turns into a law once it has been proven to be true.

    a. I agree with this statement.

    b. I disagree with this statement.

Explain your answer, using an example from your investigation about the relationship between the mass of an object and the acceleration due to gravity during free fall.

5. *Data* and *evidence* are terms that have the same meaning in science.

   a. I agree with this statement.
   b. I disagree with this statement.

   Explain your answer, using an example from your investigation about the relationship between the mass of an object and the acceleration due to gravity during free fall.

6. Why is it useful to understand the factors that control rates of change during an investigation? In your answer, be sure to include examples from at least two different investigations.

7. Why is it useful to identify patterns during an investigation? In your answer, be sure to include examples from at least two different investigations.

Lab Handout

Lab 3. Projectile Motion: How Do Changes to the Launch Angle, the Initial Velocity, and the Mass of a Projectile Affect Its Hang Time?

Introduction

Projectile motion is defined as the flight of an object near the Earth's surface under the action of gravity alone. Understanding the factors that affect the motion of projectiles has led to several major milestones in human history. Take spears as an example. Anthropologists have found evidence that human ancestors used stone-tipped spears for hunting 500,000 years ago (Wilkins et al. 2012). The ability to make and then use a spear to hunt large animals allowed our ancestors to gather more food. Later on, as people developed a better understanding of the factors that govern projectile motion, they were able to build new tools for launching projectiles that could travel farther in the air and hit targets with great accuracy. These tools included such things as bows and arrows, trebuchets, and cannons.

FIGURE L3.1

Projectile parabola

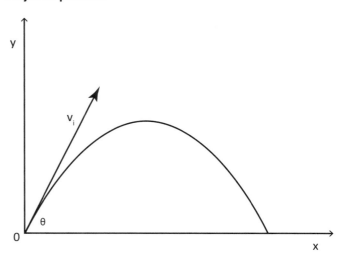

When a projectile is in flight, we assume that gravity is the sole force acting on it. The path of a projectile is a curve called a parabola, as shown in Figure L3.1. The projectile will have both a horizontal and vertical component to its velocity at any given time. Although scientists recognize that air resistance does affect the flight of a projectile, under most circumstances the effect of air resistance can be ignored. When we ignore air resistance, the initial velocity of the projectile governs the horizontal component of the projectile's velocity and the force of gravity governs its vertical component.

People often want to be able to predict how long a projectile will stay in the air (i.e., the hang time) after it is launched. There are a number of variables that may, or may not, affect the hang time of a projectile. These variables include the launch angle (denoted as θ) the initial velocity of the projectile, and the mass of the projectile. Some of these variables may also interact with each other, so the effect of any one variable may differ depending on the value of another variable. People therefore need to understand not only how these three variables affect the motion of a projectile but also how they interact with each other to predict how long a projectile will remain in the air after it is launched.

LAB 3

Your Task

Use what you to know about both linear and projectile motion, the importance of looking for patterns in data, and how to identify causal relationships to design and carry out a series of experiments to determine which variables impact the hang time of a projectile.

The guiding question of this investigation is, *How do changes to the launch angle, the initial velocity, and the mass of a projectile affect its hang time?*

Materials

You may use any of the following materials during your investigation:

Consumable	Equipment
• Tape	• Safety glasses or goggles (required)
	• Marble launcher
	• Marbles (different sizes and masses)
	• Protractor
	• Tape measure
	• Stopwatch

If you have access to the following equipment, you may also consider using a video camera and a computer or tablet with video analysis software.

Safety Precautions

Follow all normal lab safety rules. In addition, take the following safety precautions:

1. Wear sanitized safety glasses or goggles during lab setup, hands-on activity, and takedown.

2. Do not throw marbles or launch them at people.

3. Do not stand on tables or chairs.

4. Pick up marbles and other materials on the floor immediately to avoid a trip, slip, or fall hazard.

5. Wash hands with soap and water after completing the lab.

Investigation Proposal Required?   ☐ Yes   ☐ No

Getting Started

To answer the guiding question, you will need to design and carry out several different experiments. Each experiment should look at one potential variable that may or may not affect the hang time of a projectile. To accomplish this task, you must determine what type of data you need to collect, how you will collect it, and how you will analyze it.

To determine *what type of data you need to collect,* think about the following questions:

- What are the boundaries and the components of the system you are studying?
- Which variable or variables could cause a change in the hang time of a projectile?
- What information will you need to track changes in these variables?
- What information will you need to be able to track changes in the hang time of a projectile?
- What will be the independent variable and the dependent variable for each experiment?

To determine *how you will collect the data,* think about the following questions:

- What conditions need to be satisfied to establish a cause-and-effect relationship?
- What will you hold constant during each experiment?
- What variables will need to be controlled during each experiment, and how will you control them?
- How will you make sure that your data are of high quality (i.e., how will you reduce error)?

To determine *how you will analyze the data,* think about the following questions:

- What types of patterns might you look for as you analyze your data?
- How could you use mathematics to describe a relationship between variables?
- What type of calculations will you need to make?

Connections to the Nature of Scientific Knowledge and Scientific Inquiry

As you work through your investigation, you may want to consider

- the difference between observations and inferences in science, and
- the nature and role of experiments in science.

Initial Argument

Once your group has finished collecting and analyzing your data, your group will need to develop an initial argument. Your initial argument needs to include a claim, evidence to support your claim, and a justification of the evidence. The *claim* is your group's answer to the guiding question. The *evidence* is an analysis and interpretation of your data. Finally, the *justification* of the evidence is why your group thinks the evidence matters. The justification of the evidence is important because scientists can use different kinds of evidence to support their claims. Your group will create your initial argument on a whiteboard. Your whiteboard should include all the information shown in Figure L3.2 (p. 36).

LAB 3

Argumentation Session

The argumentation session allows all of the groups to share their arguments. One or two members of each group will stay at the lab station to share that group's argument, while the other members of the group go to the other lab stations to listen to and critique the other arguments. This is similar to what scientists do when they propose, support, evaluate, and refine new ideas during a poster session at a conference. If you are presenting your group's argument, your goal is to share your ideas and answer questions. You should also keep a record of the critiques and suggestions made by your classmates so you can use this feedback to make your initial argument stronger. You can keep track of specific critiques and suggestions for improvement that your classmates mention in the space below.

Critiques about our initial argument and suggestions for improvement:

FIGURE L3.2

Argument presentation on a whiteboard

The Guiding Question:	
Our Claim:	
Our Evidence:	Our Justification of the Evidence:

If you are critiquing your classmates' arguments, your goal is to look for mistakes in their arguments and offer suggestions for improvement so these mistakes can be fixed. You should look for ways to make your initial argument stronger by looking for things that the other groups did well. You can keep track of interesting ideas that you see and hear during the argumentation in the space below. You can also use this space to keep track of any questions that you will need to discuss with your team.

Interesting ideas from other groups or questions to take back to my group:

Once the argumentation session is complete, you will have a chance to meet with your group and revise your initial argument. Your group might need to gather more data or design a way to test one or more alternative claims as part of this process. Remember, your goal at this stage of the investigation is to develop the best argument possible.

Report

Once you have completed your research, you will need to prepare an investigation report that consists of three sections. Each section should provide an answer to the following questions:

1. What question were you trying to answer and why?

2. What did you do to answer your question and why?

3. What is your argument?

Your report should answer these questions in two pages or less. This report must be typed, and any diagrams, figures, or tables should be embedded into the document. Be sure to write in a persuasive style; you are trying to convince others that your claim is acceptable or valid!

Reference

Wilkins, J., B. J. Schoville, K. S. Brown, and M. Chazan. 2012. Evidence for early hafted hunting technology. *Science* 338 (6109): 942–946.

LAB 3

Lab 3. Projectile Motion: How Do Changes to the Launch Angle, the Initial Velocity, and the Mass of a Projectile Affect Its Hang Time?

1. Given the models you created during your investigation of $t(\mathbf{v}_0)$ and $t(\theta)$, choose two instances for $\mathbf{v}_0$ and θ for which the hang time of the projectile is 2 seconds.

2. Let the acceleration due to gravity be expressed as $\mathbf{a} = -\mathbf{g}$ and the initial velocity of the projectile in the $\mathbf{y}$ direction as $\mathbf{v}_0 = \mathbf{v}\sin\theta$. Given the equation for the change in $\mathbf{y}$ below,

$$\Delta\mathbf{y} = \mathbf{v}_0 t + \frac{1}{2}\mathbf{a}t^2$$

write an equation for hang time, t, as a function of $\mathbf{v}_0$, θ, and $\mathbf{g}$.

3. Current scientific knowledge and the perspectives of individual scientists influence inferences but not observations.

   a. I agree with this statement.
   b. I disagree with this statement.

   Explain your answer, using an example from your investigation of projectile motion.

4. Scientists use experiments to prove ideas.

    a. I agree with this statement.
    b. I disagree with this statement.

    Explain your answer, using an example from your investigation of projectile motion.

5. Why is it useful to identify patterns during an investigation? In your answer, be sure to include examples from at least two different investigations.

6. Why is identifying cause-and-effect relationships so important in science? In your answer, be sure to include examples from at least two different investigations.

Application Lab

LAB 4

Lab Handout

Lab 4. The Coriolis Effect: How Do the Direction and Rate of Rotation of a Spinning Surface Affect the Path of an Object Moving Across That Surface?

Introduction

When studying the motion of objects, one of the assumptions that we often make is that the ground underneath the object being studied is stationary. That is, the frame of reference being used to study the object is not moving. It turns out, however, that very few reference frames are perfectly stationary (or, more formally, inertial). For example, imagine that a person is standing on a flatbed train car. The train is moving on a track to the east with a constant velocity. Both the person and train, as a result, have the same velocity in the horizontal direction. Now imagine that the person standing on the flatbed train car were to drop a ball while the train was moving. The person who dropped the ball would see the ball fall straight down to the ground because that person has the same reference frame as the train. A second person that is watching the train go by, however,

FIGURE L4.1

A point on the solid circle travels a farther distance in one day than a point on the dashed circle. The tangential velocity is therefore greater on the solid circle than on the dashed circle.

would see something different. That person would see the same ball fall to the ground following a curved path. This is because the ball has both a vertical and horizontal component to its velocity as viewed from the reference frame of the person standing on the ground watching the train pass.

A French mathematician and physicist, Gaspard-Gustave de Coriolis, conducted numerous investigations in the 1800s to understand the movement of various bodies when the frame of reference was rotating. Initially, his studies were conducted against a rotating disc. Subsequent work by other scientists has applied his findings, known as the Coriolis effect, to the rotation of spheres, such as Earth. One of the interesting things that scientists have found is that while the rate of rotation is the same at all points on the globe, the tangential velocity is different depending on the latitude. Although every point makes one full rotation in one day, the radius of the disk traced by a point on the globe differs based on the latitude. Points near to the poles travel a smaller circle than points near the equator. Figure L4.1 illustrates this fact. What this means is that when an airplane travels with a north-south component, the place it takes off from will have a different east-west velocity than the place it intends to

The Coriolis Effect

How Do the Direction and Rate of Rotation of a Spinning Surface Affect the Path of an Object Moving Across That Surface?

land. It is therefore important to take into account the Earth's rotation when planning a flight path between two airports.

In this investigation you will have an opportunity to explore the motion of a marble as it moves across a rotating surface. Your goal will be to determine the path the marble will take when it is rolled in the y direction across a surface that is rotating in either a clockwise or counterclockwise direction. Much like the example of the person standing on a train that is moving, the rotation of the surface on which movement takes place can lead different observers to see different paths of motion based on their frame of reference. An observer on the rotating surface (formally, a non-inertial reference frame) will see the marble take a different path than a stationary person watching from outside the rotating surface (formally, an inertial reference frame). This is because, relative to each observer, the marble has different velocity vectors. To the stationary person, the marble has two velocity components, one in the y direction and one in the x direction that is due to the rotation of the surface. To the observer on the rotating platform, however, the marble only has a velocity component in the y direction.

Your Task

Use what you know about vector quantities, measurement and scale, and systems and system models to design several experiments to determine how the motion of a rotating platform affects the movement of a marble rolled across the platform as viewed from different frames of reference.

The guiding question of this investigation is, *How do the direction and rate of rotation of a spinning surface affect the path of an object moving across that surface?*

Materials

You may use any of the following materials during your investigation:

- Safety glasses or goggles (required)
- Video camera
- Computer or tablet with video analysis software
- Rotating "Coriolis" platform
- Marble
- Stopwatch
- Ruler

Safety Precautions

Follow all normal lab safety rules. In addition, take the following safety precautions:

1. Wear sanitized safety glasses or goggles during lab setup, hands-on activity, and takedown.

2. Keep fingers and toes out of the way of moving objects.

3. Do not throw marbles.

LAB 4

4. Pick up marbles or other materials on the floor immediately to avoid a trip, slip, or fall hazard.

5. Wash hands with soap and water after completing the lab.

Investigation Proposal Required?   ☐ Yes   ☐ No

Getting Started

To answer the guiding question, you will need to design and carry out at least two different experiments because a rotating platform can move in different directions (clockwise or counterclockwise) and at different rates. You will need to examine how both of these factors affect the movement of a marble rolled across the platform as viewed from different frames of reference. As you design your two experiments, you must decide what type of data you need to collect, how you will collect it, and how you will analyze it.

To determine *what type of data you need to collect*, think about the following questions:

- What are the boundaries and components of the system you are studying?
- How can you describe the components of the system quantitatively?
- What will be the independent variable and the dependent variable for each experiment?
- How will you quantify a change in the independent variable?
- Which variables are vector quantities, and which variables are scalar quantities?

To determine *how you will collect the data*, think about the following questions:

- How will you define the two reference frames?
- How will you vary the rotation rate of the platform?
- How will you measure the dependent variable?
- How will you measure the magnitude and direction of any vector quantities?
- What equipment will you need to take your measurements?
- How will you make sure that your data are of high quality (i.e., how will you reduce error)?
- How will you keep track of and organize the data you collect?

To determine *how you will analyze the data*, think about the following questions:

- What types of patterns could you look for in your data?
- How could you use mathematics to describe a relationship between variables?
- What type of calculations will you need to make?

- How will you determine whether the movement of a marble rolled across the platform as viewed from different frames of reference is the same or different?
- How will you model the motion of the marble as it moves across the rotating platform?

Connections to the Nature of Scientific Knowledge and Scientific Inquiry

As you work through your investigation, you may want to consider

- the difference between data and evidence in science, and
- the role of imagination and creativity in science.

Initial Argument

Once your group has finished collecting and analyzing your data, your group will need to develop an initial argument. Your argument needs to include a claim, evidence to support your claim, and a justification of the evidence. The *claim* is your group's answer to the guiding question. The *evidence* is an analysis and interpretation of your data. Finally, the *justification* of the evidence is why your group thinks the evidence matters. The justification of the evidence is important because scientists can use different kinds of evidence to support their claims. Your group will create your initial argument on a whiteboard. Your whiteboard should include all the information shown in Figure L4.2.

FIGURE L4.2

Argument presentation on a whiteboard

The Guiding Question:	
Our Claim:	
Our Evidence:	Our Justification of the Evidence:

Argumentation Session

The argumentation session allows all of the groups to share their arguments. One or two members of each group will stay at the lab station to share that group's argument, while the other members of the group go to the other lab stations to listen to and critique the other arguments. This is similar to what scientists do when they propose, support, evaluate, and refine new ideas during a poster session at a conference. If you are presenting your group's argument, your goal is to share your ideas and answer questions. You should also keep a record of the critiques and suggestions made by your classmates so you can use this feedback to make your initial argument stronger. You can keep track of specific critiques and suggestions for improvement that your classmates mention in the space below.

LAB 4

Critiques about our initial argument and suggestions for improvement:

If you are critiquing your classmates' arguments, your goal is to look for mistakes in their arguments and offer suggestions for improvement so these mistakes can be fixed. You should look for ways to make your initial argument stronger by looking for things that the other groups did well. You can keep track of interesting ideas that you see and hear during the argumentation in the space below. You can also use this space to keep track of any questions that you will need to discuss with your team.

Interesting ideas from other groups or questions to take back to my group:

Once the argumentation session is complete, you will have a chance to meet with your group and revise your initial argument. Your group might need to gather more data or design a way to test one or more alternative claims as part of this process. Remember, your goal at this stage of the investigation is to develop the best argument possible.

The Coriolis Effect

How Do the Direction and Rate of Rotation of a Spinning Surface Affect the Path of an Object Moving Across That Surface?

Report

Once you have completed your research, you will need to prepare an *investigation report* that consists of three sections. Each section should provide an answer to the following questions:

1. What question were you trying to answer and why?

2. What did you do to answer your question and why?

3. What is your argument?

Your report should answer these questions in two pages or less. This report must be typed, and any diagrams, figures, or tables should be embedded into the document. Be sure to write in a persuasive style; you are trying to convince others that your claim is acceptable or valid!

LAB 4

Lab 4. The Coriolis Effect: How Do the Direction and Rate of Rotation of a Spinning Surface Affect the Path of an Object Moving Across That Surface?

Use the figure on the right to answer questions 1 and 2.

1. When viewing the motion of the Earth from above the North Pole, the Earth appears to rotate counterclockwise. A plane is traveling from Miami, Florida, to Pittsburgh, Pennsylvania. Both cities are located at approximately 80° W longitude. If the plane takes off and continues in a straight line, how will its path appear to an observer in Pittsburgh?

    a. The path the plane takes will pass directly through Pittsburgh.

    b. The path the plane takes will pass to the east of Pittsburgh.

    c. The path the plane takes will pass to the west of Pittsburgh.

    How do you know?

Map of North America

2. A plane is traveling due west from Philadelphia to San Francisco. Both cities are on almost the same latitude. Will the plane be subject to the Coriolis effect?

    a. Yes

    b. No

Explain why or why not.

3. Scientists use their imagination to help them plan investigations and to analyze the results.

   a. I agree with this statement.
   b. I disagree with this statement.

   Explain your answer, using an example from your investigation of the Coriolis effect.

4. There is a difference between data and evidence in science.

   a. I agree with this statement.
   b. I disagree with this statement.

   Explain your answer, using an example from your investigation of the Coriolis effect.

5. Why is it important for scientists to think about issues related to quantity and scale as they plan or carry out an investigation? In your answer, be sure to include examples from at least two different investigations.

6. Why do scientists use models to understand and explain complex systems? In your answer, be sure to include examples from at least two different investigations.

SECTION 3
Forces and Motion

Dynamics

Introduction Labs

Lab Handout

Lab 5. Force, Mass, and Acceleration: What Is the Mathematical Relationship Among the Net Force Exerted on an Object, the Object's Inertial Mass, and Its Acceleration?

Introduction

Western scientific thought was dominated by Aristotle's views on physics for thousands of years. According to Aristotle, all objects in the universe were made of four elements: earth, water, fire, and air. Furthermore, all objects were believed to have both a *natural motion* and a *forced motion*. The natural motion of an element was toward the center of the universe (which, in Aristotle's view, was at the center of the Earth). Not all of the elements, however, had the same degree of natural motion. The elements of earth and water can move close to the center of the universe more easily than the elements of fire and air can. The natural motion of any object, as a result, depended on the relative amount of each element in it. For example, an object made up of half earth and half water would move closer to the center of the universe than an object made up of half earth and half air. The Earth exists, according to Aristotle, because objects made up of "earth" have come to rest as close as possible to the center of the universe. Forced motion, in contrast, was all the other types of motion that could not be described by natural motion. Later theorists elaborated on Aristotle's views and introduced the concept of *impetus*, the property of an object that created forced motion. When an object lost its impetus, it would then move with natural motion toward the center of the universe. In this view, when a ball is thrown in the air, the person throwing it "imparts impetus to the ball." When the ball runs out of impetus, its natural motion causes it to fall back down.

Beginning in the late Middle Ages, several scientists, including Galileo, Copernicus, and Newton, began to question Aristotle's explanation for how objects move. Galileo, for example, demonstrated that objects fall at the same rate independent of their mass, which was contrary to Aristotle's claims that heavy objects fall faster than lighter ones due to their natural motion. Galileo also observed moons orbiting Jupiter, a phenomenon that could not be explained using Aristotle's ideas. Copernicus later claimed that the Sun, and not the Earth, was at the center of the solar system. This claim, along with Galileo's observations of the moons of Jupiter, directly contradicted Aristotle's idea of natural motion because the Earth was no longer viewed as being located at the center of the universe. Isaac Newton delivered the final blow to Aristotle's explanation about how and why things move in his book *Philosophiae Naturalis Principia Mathematica* (commonly known as the *Principia*), which was published in 1687. In the *Principia*, Newton discarded natural and forced motion as an explanatory framework and introduced three basic laws of motion. *Newton's first law*, as it has become known, states that absent a net force acting on an object, an object at rest will

Force, Mass, and Acceleration
What Is the Mathematical Relationship Among the Net Force Exerted on an Object,
the Object's Inertial Mass, and Its Acceleration?

stay at rest and an object in motion will move in a straight line with constant velocity. This law also means that when a net force acts on an object, it will cause it to change how it is moving. In other words, a net force on an object will cause that object to accelerate. This law is so important that the unit for force is called a newton (N).

There are several important implications of Newton's first law. First, the idea of a net force is important, because Newton realized that multiple forces could act on an object at once. The net force, according to Newton, is just the sum of all the forces with respect to their direction. Newton showed that pushing an object to the right with a force of 10 N and pulling the object to the right with a force of 5 N is the same as if only one force of 15 N was moving the object to the right. Figure L5.1 shows another implication of Newton's first law. In this case, two people are each pushing on the box with a force of 25 N, but in opposite directions. According to Newton's first law, a push of 25 N to the left and a push of 25 N to the right results in a net force of zero (0 N). The final implication of Newton's first law is that the net force and the resulting change in motion must be in the same direction. For example, pushing an object with a force of 10 N to the right will cause it to accelerate to the right, and not upward toward the sky. Acceleration is equal to the rate of change of velocity with time, and velocity is equal to the rate of change of position with time.

FIGURE L5.1

Two people pushing on a box with equal forces but in opposite directions

In the *Principia,* Newton also provided a mathematical relationship among the net force acting on an object, that object's mass, and the acceleration of the object. This mathematical relationship is now known as Newton's second law. This mathematical relationship was revolutionary at the time because it not only explained how objects move but also allowed for people to predict changes in an object's motion. In fact, physicists still use this mathematical relationship today to predict the motion of an object when a force acts on it and to determine the amount of force needed to move heavy objects.

Your Task

Use what you know about forces and motion, causal relationships, and the importance of scales, proportions, and quantity in science to design and carry out an investigation that will allow you to determine the mathematical relationship among the net force acting on an object, its mass, and its acceleration.

The guiding question of this investigation is, *What is the mathematical relationship among the net force exerted on an object, the object's inertial mass, and its acceleration?*

Materials

You may use any of the following materials during your investigation:

Consumables
- String or fishing line
- Tape

Equipment
- Safety glasses or goggles (required)
- Dynamics cart
- Dynamics track
- Set of masses
- Electronic or triple beam balance

- 2 Pulleys
- Stopwatch
- Ruler
- Meterstick

If you have access to the following equipment, you may also consider using a video camera and computer or tablet with video analysis software.

Safety Precautions

Follow all normal lab safety rules. In addition, take the following safety precautions:

1. Wear sanitized safety glasses or goggles during lab setup, hands-on activity, and takedown.

2. Keep fingers and toes out of the way of moving objects.

3. Wash hands with soap and water after completing the lab.

Investigation Proposal Required?   ☐ Yes   ☐ No

Getting Started

To answer the guiding question, you will need to design and carry out at least two different experiments. You will need to first determine how changing the force acting on an object of a constant mass affects the acceleration of the object. Then you will need to determine how applying the same force to an object of different mass affects the acceleration of the object. You can conduct these two experiments using a cart-and-track system; Figure L5.2 shows the basic setup of the system.

FIGURE L5.2

Cart-and-track system that can be used to determine the mathematical relationship among the net force acting on an object, its mass, and its acceleration

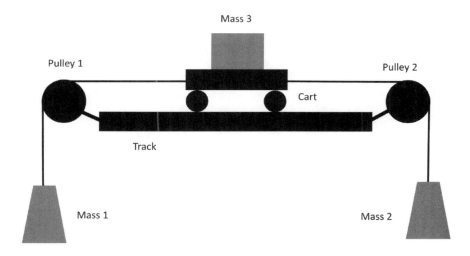

Before you can design and carry out these two experiments, you will need to decide what type of data you need to collect, how you will collect it, and how you will analyze it.

To determine *what type of data you need to collect*, think about the following questions:

- What are the boundaries and components of the system you are studying?
- How do the components of the system interact with each other?
- Which factors might control the rate of change in this system?
- How could you keep track of changes in this system quantitatively?
- What information will you need to be able to determine the acceleration of the cart?
- How will you measure the net force acting on the cart?
- What will be the independent variable and the dependent variable for each experiment?

To determine *how you will collect the data*, think about the following questions:

- How will you change the mass of the cart?
- How will you change the net force acting on the cart?
- What equipment will you need to collect the data?
- What conditions need to be satisfied to establish a cause-and-effect relationship?
- How will you measure the magnitude and direction of any vector quantities?

- How will you make sure that your data are of high quality (i.e., how will you reduce error)?
- How will you keep track of and organize the data you collect?

To determine *how you will analyze the data*, think about the following questions:

- How could you use mathematics to describe a relationship between variables?
- What type of calculations will you need to make?
- What types of graphs and equations signify a proportional relationship?
- What type of table or graph could you create to help make sense of your data?

Once you have carried out your experiments, your group will need to develop a mathematical function that you can use to predict the acceleration of an object based on its mass and the net force acting on it. The last step in this investigation will be to test your function. To accomplish this goal, you can use a different amount of mass on the cart and a different net force to determine if your function leads to accurate predictions about acceleration of the cart under different conditions. If you are able to use your function to make accurate predictions about the motion of the cart under different conditions, then you will be able to generate the evidence you need to convince others that the function you developed is valid.

Connections to the Nature of Scientific Knowledge and Scientific Inquiry

As you work through your investigation, you may want to consider

- how scientific knowledge changes over time, and
- the difference between laws and theories in science.

Initial Argument

Once your group has finished collecting and analyzing your data, your group will need to develop an initial argument. Your argument must include a claim, evidence to support your claim, and a justification of the evidence. The *claim* is your group's answer to the guiding question. The *evidence* is an analysis and interpretation of your data. Finally, the *justification* of the evidence is why your group thinks the evidence matters. The justification of the evidence is important because scientists can use different kinds of evidence to support their claims. Your group will create your initial argument on a whiteboard. Your whiteboard should include all the information shown in Figure L5.3.

Argumentation Session

The argumentation session allows all of the groups to share their arguments. One or two members of each group will stay at the lab station to share that group's argument, while the other members of the group go to the other lab stations to listen to and critique the other arguments. This is similar to what scientists do when they propose, support, evaluate, and refine new ideas during a poster session at a conference. If you are presenting your group's argument, your goal is to share your ideas and answer questions. You should also keep a record of the critiques and suggestions made by your classmates so you can use this feedback to make your initial argument stronger. You can keep track of specific critiques and suggestions for improvement that your classmates mention in the space below.

Critiques about our initial argument and suggestions for improvement:

FIGURE L5.3
Argument presentation on a whiteboard

The Guiding Question:	
Our Claim:	
Our Evidence:	Our Justification of the Evidence:

If you are critiquing your classmates' arguments, your goal is to look for mistakes in their arguments and offer suggestions for improvement so these mistakes can be fixed. You should look for ways to make your initial argument stronger by looking for things that the other groups did well. You can keep track of interesting ideas that you see and hear during the argumentation in the space below. You can also use this space to keep track of any questions that you will need to discuss with your team.

Interesting ideas from other groups or questions to take back to my group:

Once the argumentation session is complete, you will have a chance to meet with your group and revise your initial argument. Your group might need to gather more data or design a way to test one or more alternative claims as part of this process. Remember, your goal at this stage of the investigation is to develop the best argument possible.

Report

Once you have completed your research, you will need to prepare an investigation report that consists of three sections. Each section should provide an answer to the following questions:

1. What question were you trying to answer and why?

2. What did you do to answer your question and why?

3. What is your argument?

Your report should answer these questions in two pages or less. This report must be typed, and any diagrams, figures, or tables should be embedded into the document. Be sure to write in a persuasive style; you are trying to convince others that your claim is acceptable or valid!

Reference

Newton, I. 1687. *Philosophiae naturalis principia mathematica* [Mathematical principles of natural philosophy]. London: S. Pepys.

Checkout Questions

Lab 5. Force, Mass and Acceleration: What Is the Mathematical Relationship Among the Net Force Exerted on an Object, the Object's Inertial Mass, and Its Acceleration?

1. In mathematics, two variables are proportional if a change in one variable is always accompanied by a change in another variable. Which, if any, variables from your investigation are proportional? You may choose more than one answer.

    a. Force and mass

    b. Force and acceleration

    c. Mass and acceleration

    What evidence do you have to support your claim?

Use the following information to answer questions 2–4. Two people are playing a game of tug-of-war with the rope attached to a mass of 25 kg at the center. The person pulling to the left pulls with a force of 20 N. The person pulling to the right pulls with a force of 10 N.

2. Which direction will the 25 kg mass move?

    a. Left

    b. Right

    c. It will not move

    How do you know?

3. What will the velocity of the mass be after 1 second?

4. What will the velocity of the mass be after 2 seconds?

5. Two high school physics students are talking, and one says that an acceleration can cause a force, while the other says that a force causes an object to accelerate.

    a. I agree with the first person.
    b. I agree with the second person.

   Use the concept of a cause-and-effect relationship to explain your answer.

6. Force is directly proportional to both mass and acceleration.

    a. I agree with this statement.
    b. I disagree with this statement.

   Use the concept of a proportional relationship to explain your answer.

Force, Mass, and Acceleration
*What Is the Mathematical Relationship Among the Net Force Exerted on an Object,
the Object's Inertial Mass, and Its Acceleration?*

7. Aristotle's views about physics dominated scientific thought for thousands of years. Yet, ideas from Galileo, Newton, and others are now accepted scientific laws and theories. In other words, the accepted "scientific view" of the world changed. Explain what led others to discard Aristotle's views for Galileo's and Newton's, and why it is important for scientists to be open to differing ideas, using an example from your investigation about force, mass, and acceleration.

8. In science, there is a distinction between a law and a theory. What makes laws and theories different in science, and why does this distinction matter? Include an example from your investigation about force, mass, and acceleration in your answer.

LAB 6

Lab Handout

Lab 6. Forces on a Pulley: How Does the Mass of the Counterweight Affect the Acceleration of a Pulley System?

Introduction

There are many products that we now take for granted but were revolutionary when they were first invented. One of these products is the elevator. When elevators were first invented, there were numerous design problems related to moving the car up and down an elevator shaft. One of these problems was controlling the acceleration of the car as the mass of the car changed with the addition of people. In 1861, Elisha G. Otis patented a new design for an elevator; Figure L6.1 shows the patent drawing. To help moderate the acceleration of the car, Otis added a counterweight to the system. The counterweight is attached to the elevator via two pulleys at the top of the elevator shaft. This basic idea is still used in elevators today.

FIGURE L6.1

Drawings from Otis's 1861 patent application for an elevator

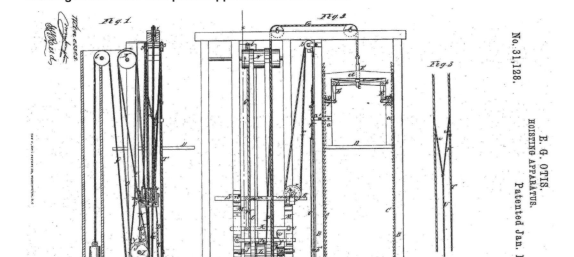

The basis for Otis's decision to add the counterweight to the system came from Newton's second law of motion, which states that the acceleration of a system is proportional to the net forces acting on the system. Acceleration is equal to the rate of change of velocity with time, and velocity is equal to the rate of change of position with time. A counterweight works because the force of gravity pulls both the elevator car down and the counterweight down. Because the counterweight is on a different side of the pulley from the elevator car, the force of gravity on the counterweight acts in a different direction (relative to the pulley) than the force of gravity acting on the elevator car. Thus, the total acceleration of the elevator car and counterweight, in accordance with Newton's second law of motion, is equal to the sum of the forces acting on each part of the system divided by the total mass of the system.

Otis's elevator design is an application of one class of simple machines called pulleys. The first scientist to study simple machines was Archimedes. Although he did some research on the behavior of pulleys, his main focus was on the behavior of a simple machine called the screw. Renaissance scientists took up the mantle from Archimedes and began to explore the behavior of other simple machines, including wheels, levers, and pulleys. However, it was not until the work of Isaac Newton that scientists were able to mathematically model the motion of simple machines. This ability to use equations to describe the forces acting on a simple machine allowed physicists and engineers to apply these equations to the design of new systems that incorporated much more complicated machines. Otis, for example, applied the mathematical insights of Newton's laws to study pulleys that could be used to improve on the design of an elevator.

Your Task

Use what you know about forces and motion, systems and system models, stability and change in systems, and the relationship between structure and function in designed systems to design and carry out an investigation to explore the relationship between the mass of the counterweight and the acceleration of the pulley system. You will then develop a conceptual model that can be used to explain the behavior of this simple machine. Once you have developed your model, you will need to test it to determine if it allows you to make accurate predictions about its behavior.

The guiding question of this investigation is, *How does the mass of the counterweight affect the acceleration of a pulley system?*

Materials

You may use any of the following materials during your investigation:

- Safety glasses or goggles (required)
- Ring stand
- Pulley
- Hanging mass set
- Electronic or triple beam balance
- String or fishing line
- Stopwatch
- Ruler
- Meterstick

If you have access to the following equipment, you may also consider using a video camera and a computer or tablet with video analysis software.

LAB 6

Safety Precautions

Follow all normal lab safety rules. In addition, take the following safety precautions:

1. Wear sanitized safety glasses or goggles during lab setup, hands-on activity, and takedown.

2. Keep your fingers and toes out of the way of moving objects.

3. Wash hands with soap and water after completing the lab.

Investigation Proposal Required?   ☐ Yes   ☐ No

Getting Started

The first step in developing your model is to determine how the mass of a counterweight affects the acceleration of a pulley system. One way to gather the information you need is to create a physical model of the system to see how it behaves. In this investigation, you can use a set of hanging masses and a pulley attached to a support stand to study the movement of the entire system. Figure L6.2 shows how you can set up a pulley system using the available materials. Before you can design your investigation, however, you must determine what type of data you need to collect, how you will collect it, and how you will analyze it.

To determine *what type of data you need to collect,* think about the following questions:

- What are the boundaries and components of the system you are studying?
- How might the structure of the pulley system determine its function?
- Which factor(s) might control the rate of change in this system?
- How could you keep track of changes in this system quantitatively?
- What forces are acting on the system?
- What forces act on each mass?
- How will you determine the acceleration of the system?
- What will be the independent variable and the dependent variable?

To determine *how you will collect the data,* think about the following questions:

FIGURE L6.2

Equipment used to explore the movement of a pulley system

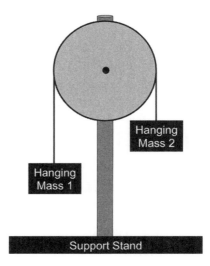

- What other factors do you need to control for as you collect data?
- What measurement scale or scales should you use to collect data?
- What equipment will you need to take the measurements?
- For any vector quantities, which directions will be positive and which directions will be negative?
- How will you make sure that your data are of high quality (i.e., how will you reduce error)?
- How will you keep track of and organize the data you collect?

To determine *how you will analyze the data,* think about the following questions:

- How could you use mathematics to describe a change over time?
- How could you use mathematics to describe a relationship between variables?
- What type of calculations will you need to make?
- What will be the positive convention and what will be the negative convention?

Once you have determined how the mass of the counterweight affects the acceleration of a pulley system, your group will need to develop a conceptual model to explain it. Your model also must include information about all the forces acting on each component or structure found in the entire pulley system.

The last step in this investigation is to test your model. To accomplish this goal, you can use different hanging masses (ones that you did not use earlier) to determine if your model enables you to make accurate predictions about the acceleration of the entire pulley system. If you are able to use your model to make accurate predictions about the function of the system based on the structure of that system, then you will be able to generate the evidence you need to convince others that your model is valid.

Connections to the Nature of Scientific Knowledge and Scientific Inquiry

As you work through your investigation, you may want to consider

- the difference between observations and inferences in science, and
- how the culture of science, societal needs, and current events influence the work of scientists.

Initial Argument

Once your group has finished collecting and analyzing your data, your group will need to develop an initial argument. Your argument must include a claim, evidence to support your claim, and a justification of the evidence. The *claim* is your group's answer to the guiding question. The *evidence* is an analysis and interpretation of your data. Finally, the

LAB 6

justification of the evidence is why your group thinks the evidence matters. The justification of the evidence is important because scientists can use different kinds of evidence to support their claims. Your group will create your initial argument on a whiteboard. Your whiteboard should include all the information shown in Figure L6.3.

Argumentation Session

The argumentation session allows all of the groups to share their arguments. One or two members of each group will stay at the lab station to share that group's argument, while the other members of the group go to the other lab stations to listen to and critique the other arguments. This is similar to what scientists do when they propose, support, evaluate, and refine new ideas during a poster session at a conference. If you are presenting your group's argument, your goal is to share your ideas and answer questions. You should also keep a record of the critiques and suggestions made by your classmates so you can use this feedback to make your initial argument stronger. You can keep track of specific critiques and suggestions for improvement that your classmates mention in the space below.

Critiques about our initial argument and suggestions for improvement:

FIGURE L6.3

Argument presentation on a whiteboard

The Guiding Question:	
Our Claim:	
Our Evidence:	Our Justification of the Evidence:

If you are critiquing your classmates' arguments, your goal is to look for mistakes in their arguments and offer suggestions for improvement so these mistakes can be fixed. You should look for ways to make your initial argument stronger by looking for things that the other groups did well. You can keep track of interesting ideas that you see and hear during the argumentation in the space below. You can also use this space to keep track of any questions that you will need to discuss with your team.

Interesting ideas from other groups or questions to take back to my group:

Once the argumentation session is complete, you will have a chance to meet with your group and revise your initial argument. Your group might need to gather more data or design a way to test one or more alternative claims as part of this process. Remember, your goal at this stage of the investigation is to develop the best argument possible.

Report

Once you have completed your research, you will need to prepare an investigation report that consists of three sections. Each section should provide an answer to the following questions:

1. What question were you trying to answer and why?

2. What did you do to answer your question and why?

3. What is your argument?

Your report should answer these questions in two pages or less. This report must be typed, and any diagrams, figures, or tables should be embedded into the document. Be sure to write in a persuasive style; you are trying to convince others that your claim is acceptable or valid!

LAB 6

Lab 6. Forces on a Pulley: How Does the Mass of the Counterweight Affect the Acceleration of a Pulley System?

Use the information in the figure at right and Newton's laws of motion to answer questions 1 and 2. In the figure, assume that the 5 g mass is the counterweight.

1. What is the tension in the string connecting the two masses?

   How do you know?

5 g

25 g

Support Stand

2. Is the tension in the string constant, or does it change depending on the mass of the counterweight? In answering this question, assume the string and the pulley are massless.

3. In physics, it is important to define the system under study in order to understand how objects move.

   a. I agree with this statement.

   b. I disagree with this statement.

Explain your answer, using an example from your investigation about a pulley system.

4. How a system is structured influences how that system functions.

   a. I agree with this statement.
   b. I disagree with this statement.

   Explain your answer, using an example from your investigation about a pulley system.

5. In science, there is a distinction between observation and inference. Explain why this distinction is important, using an example from your investigation about a pulley system.

6. People view some research as being more important than other research in science because of the cultural values, societal needs, or current events. On another sheet of paper, explain why it is important to understand how the culture of science, societal needs, and current events influence the work of scientists, using an example from your investigation about a pulley system.

Lab Handout

Lab 7. Forces on an Incline: What Is the Mathematical Relationship Between the Angle of Incline and the Acceleration of an Object Down the Incline?

Introduction

Physicists have been studying the motion of objects on an incline for centuries because most of the world is not flat. This line of research is important because many people live on or near hills or mountains and must build homes, store equipment, and travel from one location to another on an incline. Engineering of large structures has required further study of how objects do or do not move down an incline. For example, when building parking garages, many engineers build garages so that cars can be parked on the incline of the garage between levels, and not just on the flat surface.

Scientists often use Newton's laws of motion to understand the relationship between the net force exerted on an object on an incline plane, its inertial mass, and its acceleration. Acceleration is equal to the rate of change of velocity with respect to time, and velocity is equal to the rate of change of position with respect to time. Gravity is one of the forces that act on an object on an incline plane. The force of gravity pulls the object toward the center of the Earth. This is important to keep in mind because it means that gravity does not act directly down the incline. Figure L7.1 shows how the force of gravity (**Fg**) acts on a box sitting on a ramp. It also shows how scientists measure the angle of incline. As can be seen in this figure, the angle of incline (θ) is measured relative to a flat, horizontal surface (as opposed to measuring the angle from the vertical). Therefore, when the measure of the angle is 0°, the surface is flat. When the angle of incline is 90°, it is a vertical surface and an object no longer moves down the surface, but instead moves in free fall.

FIGURE L7.1

A force diagram for an object on an incline plane

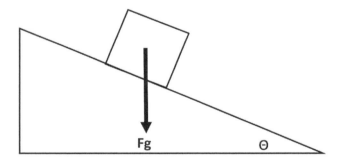

As mentioned above, it is important for us to understand how the angle of an incline plane affects the motion of an object on the incline for many different reasons. An engineer who is responsible for designing and building a parking garage, for example, needs to understand how the angle of incline in the garage will affect the motion of parked cars because if he or she makes the incline in the garage too steep, then the cars may roll or slide down the incline, even when they are in park. When this happens, property can be damaged and people can get hurt. In this investigation, you will have an opportunity to learn more about the relationship between the angle of an incline and the motion of object on that incline.

Your Task

Use what you know about forces and motion, the importance of looking for patterns, and the relationship between structure and function to design and carry out an investigation to examine how a toy car moves down the incline. The goal of this investigation is to come up with an equation that you can use to predict the acceleration of the toy car down as it moves down an incline.

The guiding question of this investigation is, **What is the mathematical relationship between the angle of incline and the acceleration of an object down the incline?**

Materials

You may use any of the following materials during your investigation:

Consumables
- Tape
- String or fishing line

Equipment
- Safety glasses or goggles (required)
- Plastic cup
- Small toy car or cart
- Electronic or triple beam balance

- Small masses
- Meterstick
- Protractor
- Ramp or board
- Stopwatch
- Pulley

If you have access to the following equipment, you may also consider using a video camera and a computer or tablet with video analysis software.

Safety Precautions

Follow all normal lab safety rules. In addition, take the following safety precautions:

1. Wear sanitized safety glasses or goggles during lab setup, hands-on activity, and takedown.

2. Keep fingers and toes out of the way of moving objects.

3. Do not stand on tables and chairs.

4. Wash hands with soap and water after completing the lab.

LAB 7

Investigation Proposal Required?   ☐ Yes   ☐ No

Getting Started

To answer the guiding question, you will need to design and carry out an experiment. To accomplish this task, you must determine what type of data you need to collect, how you will collect it, and how you will analyze it.

To determine *what type of data you need to collect*, think about the following questions:

- What are the boundaries and components of the system you are studying?
- How do the components of the system interact with each other?
- Which factors might control the rate of change in velocity of the toy car?
- How might a change in the structure of the incline affect how the moving object functions?
- How could you keep track of changes in this system quantitatively?
- What forces are acting on the toy car when it is on the incline?
- What information will you need to be able to determine the acceleration of the toy car?
- What will be the independent variable and the dependent variable for your experiment?

To determine *how you will collect the data*, think about the following questions:

- How will you change the independent variable during your experiment?
- What other factors will you need to hold constant during your experiment?
- How will you measure the magnitude and direction of any vector quantities?
- What equipment will you need to collect the data?
- How will you make sure that your data are of high quality (i.e., how will you reduce error)?
- How will you keep track of and organize the data you collect?

To determine *how you will analyze the data*, think about the following questions:

- What types of patterns might you look for as you analyze your data?
- How could you use mathematics to describe a relationship between variables?
- What type of calculations will you need to make?
- What types of graphs and equations signify a proportional relationship?
- What type of table or graph could you create to help make sense of your data?

Connections to the Nature of Scientific Knowledge and Scientific Inquiry

As you work through your investigation, you may want to consider

- the difference between observations and inferences in science, and
- how scientists use different methods to answer different types of questions.

Initial Argument

Once your group has finished collecting and analyzing your data, your group will need to develop an initial argument. Your initial argument needs to include a claim, evidence to support your claim, and a justification of the evidence. The *claim* is your group's answer to the guiding question. The *evidence* is an analysis and interpretation of your data. Finally, the *justification* of the evidence is why your group thinks the evidence matters. The justification of the evidence is important because scientists can use different kinds of evidence to support their claims. Your group will create your initial argument on a whiteboard. Your whiteboard should include all the information shown in Figure L7.2.

FIGURE L7.2

Argument presentation on a whiteboard

The Guiding Question:	
Our Claim:	
Our Evidence:	Our Justification of the Evidence:

Argumentation Session

The argumentation session allows all of the groups to share their arguments. One or two members of each group will stay at the lab station to share that group's argument, while the other members of the group go to the other lab stations to listen to and critique the other arguments. This is similar to what scientists do when they propose, support, evaluate, and refine new ideas during a poster session at a conference. If you are presenting your group's argument, your goal is to share your ideas and answer questions. You should also keep a record of the critiques and suggestions made by your classmates so you can use this feedback to make your initial argument stronger. You can keep track of specific critiques and suggestions for improvement that your classmates mention in the space below.

Critiques about our initial argument and suggestions for improvement:

LAB 7

If you are critiquing your classmates' arguments, your goal is to look for mistakes in their arguments and offer suggestions for improvement so these mistakes can be fixed. You should look for ways to make your initial argument stronger by looking for things that the other groups did well. You can keep track of interesting ideas that you see and hear during the argumentation in the space below. You can also use this space to keep track of any questions that you will need to discuss with your team.

Interesting ideas from other groups or questions to take back to my group:

Once the argumentation session is complete, you will have a chance to meet with your group and revise your initial argument. Your group might need to gather more data or design a way to test one or more alternative claims as part of this process. Remember, your goal at this stage of the investigation is to develop the best argument possible.

Report

Once you have completed your research, you will need to prepare an investigation report that consists of three sections. Each section should provide an answer to the following questions:

1. What question were you trying to answer and why?

2. What did you do to answer your question and why?

3. What is your argument?

Your report should answer these questions in two pages or less. This report must be typed, and any diagrams, figures, or tables should be embedded into the document. Be sure to write in a persuasive style; you are trying to convince others that your claim is acceptable or valid!

Checkout Questions

Lab 7. Forces on an Incline: What Is the Mathematical Relationship Between the Angle of Incline and the Acceleration of an Object Down the Incline?

1. Draw a free-body diagram of the forces acting on a toy car accelerating down an incline.

2. In your investigation, you determined the relationship between the angle of incline and the acceleration of a toy car moving down the incline. Assuming there is no friction between the incline and the object that is placed on it, would your equation change if the mass of the object were increased?

   a. Yes
   b. No

   Explain your answer.

3. The wheels on the toy car reduced the impact of the force of friction on the motion of the toy car as it traveled down the incline. What would your data have looked like if the frictional force acting on the incline were larger? Use data from your investigation as part of your answer.

4. Scientists can make different inferences from the same observations.

    a. I agree with this statement.

    b. I disagree with this statement.

Explain your answer, using an example from your investigation about objects moving down an incline.

5. There is a scientific method that all scientists must follow.

    a. I agree with this statement.

    b. I disagree with this statement.

Explain your answer, using an example from your investigation about objects moving down an incline.

6. Why is important to look for patterns in science? In your answer, be sure to include examples from at least two different investigations.

7. Why is important to think about the relationship between structure and function in nature during an investigation? In your answer, be sure to include examples from at least two different investigations.

Application Labs

LAB 8

Lab Handout

Lab 8. Friction: Why Are Some Lubricants Better Than Others at Reducing the Coefficient of Friction Between Metal Plates?

Introduction

Friction plays an important role in many of our daily experiences. In many cases, friction between two surfaces is beneficial. Friction between your shoes and the ground allows you to walk; friction between your tires and the road allows your car to move forward; and friction keeps your cell phone from sliding off a table if you accidently bump into it. Frictional forces also produce heat. When you rub your hands together on a cold day, for example, friction produces heat and warms your hands. Sometimes, however, producing heat from friction is an undesirable outcome or unwanted by-product of a specific process. Take the internal combustion engine found in most cars as an example. An internal combustion engine has a set of pistons that are constantly moving when the engine is on. Each of these pistons is found inside a metal cylinder (see Figure L8.1). For the engine to work, each piston must be able to slide up and down inside a metal cylinder. The sliding motion, however, produces heat because of the friction that exists between the piston and the cylinder, which can cause parts of the engine to heat up and eventually reach a temperature where the different parts of the engine expand and deform or break.

The coefficient of friction is the ratio of the frictional force and the normal force that exists between two surfaces when they are in contact. There are two types of friction coefficients. The first type is called the *coefficient of static friction*. This measure is used when there is no relative motion between the two surfaces. For example, when a car is at rest we describe the friction that exists between the tires and the ground using the coefficient of static friction. The second type is called the *coefficient of kinetic friction*. This measure is used when there is relative motion between two surfaces. The coefficient of kinetic friction affects the amount of heat that is generated when objects that are in contact move past each other. In general, as the coefficient of kinetic friction increases, so does the amount of heat that is produced as the result of friction. Finally, it is important to keep in mind that a coefficient of friction, whether it is static or kinetic, only exists between materials that are in contact. For example, the coefficient of kinetic friction between iron and copper is 0.29 when these two materials are in contact, and the coefficient of kinetic friction between iron and zinc is 0.21 when these two materials are in contact. Iron does not have a coefficient of kinetic friction unless it is in contact with another material, such as copper or zinc.

As mentioned earlier, when a piston slides up and down in a cylinder, friction causes the engine to heat up quickly. Engineers therefore use *lubricants* to reduce the coefficient of kinetic friction between the piston and the cylinder. A lubricant is simply a substance that is put in between two materials to reduce the heat generated when the two surfaces move past each other. Oils are often used as lubricants. An oil can be defined as a viscous liquid

Why Are Some Lubricants Better Than Others at Reducing the Coefficient of Friction Between Metal Plates?

that is composed of hydrogen, oxygen, and carbon atoms. Oils do not mix with water, and they feel slippery. *Viscosity* is a measure of how resistant a liquid is to flow. Honey, for example, has a higher viscosity than water, so honey flows slower than water. There are many different types of oils. Each one has a unique set of physical properties. Table L8.1 includes the physical properties and sources of 10 different types of oil. Some of these oils are better than others at reducing the coefficient of kinetic friction between metals because of their unique physical properties.

To be able to make new and better lubricants for engines, it is important to understand what makes oil such a good lubricant. Scientists and engineers, in other words, need to understand how the different physical properties of an oil affect or do not affect its ability to reduce the coefficient of kinetic friction between two materials. This type of research is

FIGURE L8.1 _____

A cross-section of an internal combustion engine, showing three different pistons. Each piston is found inside a cylinder.

TABLE L8.1 _____

Sources and physical properties of different types of oils

Name	Source	Physical properties			
		Density (g/cm³)	Viscosity (cP)	Melting point (°C)	Boiling point (°C)
Motor (SAE 40)	Petroleum	0.90	319	−12	315
Motor (SAE 30)	Petroleum	0.89	200	−30	300
Motor (SAE 20)	Petroleum	0.88	125	No data	280
Motor (SAE 10)	Petroleum	0.87	65	No data	260
Canola	Plants	0.91	57	−10	205
Castor	Plants	0.96	985	−18	313
Corn	Plants	0.90	81	−11	230
Mineral	Petroleum	0.87	44	−9	300
Olive	Plants	0.91	84	−6.0	300
Peanut	Plants	0.91	68	3	255

Note: cP = centipoise.

important because it is not enough to determine which type of lubricant works best as a lubricant (like a product reviewer would) if the goal of the scientist or engineer is to create a better product. Instead, scientists or engineers must understand why a lubricant works. It is also important for scientists and engineers to understand how different conditions, such as when there is a high coefficient of kinetic friction or a low coefficient of kinetic friction between two objects, affect how well a lubricant works. Your goal for this investigation is to explain why a lubricant, such as oil, is able to reduce the coefficient of kinetic friction between two surfaces.

Your Task

Use what you know about forces and motion, structure and function, and the role models of systems play in science to design and carry out an investigation to determine how different types of oils change the coefficient of kinetic friction between two metal plates. You will then use what you know about structure and function to develop a model that can be used to explain why some oils reduce the coefficient of friction more than other oils. Your model can be conceptual, mathematical, or graphical. To be valid or acceptable, your model must take into account the different physical properties of the oils. Once you have developed your model, you will need to test it to determine if it allows you to predict how the use of other oils (which you have not tested before) will change the coefficient of kinetic friction between two plates, using only the physical properties of these other oils as a guide.

The guiding question of this investigation is, *Why are some lubricants better than others at reducing the coefficient of friction between metal plates?*

Materials

You may use any of the following materials during your investigation:

Consumables	Equipment
• Motor oil (SAE 10)	• Safety glasses or goggles (required)
• Motor oil (SAE 20)	• Chemical-resistant apron (required)
• Motor oil (SAE 30)	• Gloves (required)
• Motor oil (SAE 40)	• Electronic or triple beam balance
• Canola oil	• Aluminum metal plates
• Castor oil	• Brass metal plates
• Corn Oil	• Steel metal plates
• Mineral oil	• Spring scale or force probe
• Olive oil	• Mass set
• Peanut oil	• Tray
• Wax paper	• Stopwatch
• Tape	• Meterstick
• String	

Safety Precautions

Follow all normal lab safety rules. Your teacher will explain relevant and important information about working with the chemicals associated with this investigation. In addition, take the following safety precautions:

1. Follow safety precautions noted on safety data sheets for hazardous chemicals.

2. Wear sanitized indirectly vented chemical-splash goggles and chemical-resistant, nonlatex gloves and aprons during lab setup, hands-on activity, and takedown.

3. Never put consumables in your mouth.

4. Do not eat or drink any food items used in the lab activity.

5. Consult with the teacher about disposal of waste oil. Do not pour any chemicals down the lab sink.

6. Clean up any spilled liquid on the floor immediately to avoid a slip or fall hazard.

7. Wash hands with soap and water after completing the lab.

Investigation Proposal Required?   ☐ Yes   ☐ No

Getting Started

The first step in developing your model is to determine how the addition of different types of oils changes the coefficient of kinetic friction between two metal plates. One way to gather the information you need to determine the coefficient of kinetic friction between two metal plates is to measure the amount of force it takes to pull a metal plate across another one. Figure L8.2 shows how you can measure this force.

FIGURE L8.2
Equipment used to measure the force required to pull a metal plate across another metal plate

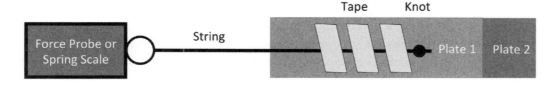

Before you can begin conducting your tests, however, you must decide what other type of data you will need to collect, how you will collect it, and how you will analyze it.

To determine *what type of data you need to collect,* think about the following questions:

- What are the boundaries and components of the system you are studying?

- How do the components of the system interact with each other?
- How could you keep track of changes in this system quantitatively?
- How might the structure of a lubricant impact its function in reducing the coefficient of friction?
- What information do you need to calculate the coefficient of kinetic friction?
- What are all the forces that are acting on each object?
- Are any of the pairs of forces balanced?

To determine *how you will collect the data,* think about the following questions:

- What comparisons will you need to make?
- What other factors will you need to control during your tests?
- How will you make sure that your data are of high quality (i.e., how will you reduce error)?
- How will you keep track of and organize the data you collect?

To determine *how you will analyze the data,* think about the following questions:

- What types of patterns could you look for in your data?
- What type of calculations will you need to make to determine a coefficient of kinetic friction?
- How could you use mathematics to describe a change in a coefficient of kinetic friction?
- How could you use mathematics to document a difference between types of lubricants?

Once you have determined how the addition of different types of oils changes the coefficient of kinetic friction between two metal plates, your group will need to develop your conceptual model. The model must be able to explain why some of the oils were better than others at reducing the coefficient of friction between metal plates. Your model also must include information about the physical properties of the oils. The physical properties of each oil are provided in Table L8.1. Your model should also include information about what you think is happening between the oils and the plates on the submicroscopic level.

The last step in this investigation is to test your model. To accomplish this goal, you can use different oils (ones that you did not test) to determine if your model enables you to make accurate predictions about how these oils will change the coefficient of kinetic friction between two metal plates. If you are able to use your model to make accurate predictions about how the oils function as a lubricant based on their structure, then you will be able to generate the evidence you need to convince others that your model is valid.

Connections to the Nature of Scientific Knowledge and Scientific Inquiry

As you work through your investigation, you may want to consider

- how the culture of science, societal needs, and current events influence the work of scientists; and
- how scientists use different methods to answer different types of questions.

Initial Argument

Once your group has finished collecting and analyzing your data, your group will need to develop an initial argument. Your initial argument needs to include a claim, evidence to support your claim, and a justification of the evidence. The *claim* is your group's answer to the guiding question. The *evidence* is an analysis and interpretation of your data. Finally, the *justification* of the evidence is why your group thinks the evidence matters. The justification of the evidence is important because scientists can use different kinds of evidence to support their claims. Your group will create your initial argument on a whiteboard. Your whiteboard should include all the information shown in Figure L8.3.

FIGURE L8.3

Argument presentation on a whiteboard

The Guiding Question:	
Our Claim:	
Our Evidence:	Our Justification of the Evidence:

Argumentation Session

The argumentation session allows all of the groups to share their arguments. One or two members of each group will stay at the lab station to share that group's argument, while the other members of the group go to the other lab stations to listen to and critique the other arguments. This is similar to what scientists do when they propose, support, evaluate, and refine new ideas during a poster session at a conference. If you are presenting your group's argument, your goal is to share your ideas and answer questions. You should also keep a record of the critiques and suggestions made by your classmates so you can use this feedback to make your initial argument stronger. You can keep track of specific critiques and suggestions for improvement that your classmates mention in the space below.

Critiques about our initial argument and suggestions for improvement:

LAB 8

If you are critiquing your classmates' arguments, your goal is to look for mistakes in their arguments and offer suggestions for improvement so these mistakes can be fixed. You should look for ways to make your initial argument stronger by looking for things that the other groups did well. You can keep track of interesting ideas that you see and hear during the argumentation in the space below. You can also use this space to keep track of any questions that you will need to discuss with your team.

Interesting ideas from other groups or questions to take back to my group:

Once the argumentation session is complete, you will have a chance to meet with your group and revise your initial argument. Your group might need to gather more data or design a way to test one or more alternative claims as part of this process. Remember, your goal at this stage of the investigation is to develop the best argument possible.

Report

Once you have completed your research, you will need to prepare an investigation report that consists of three sections. Each section should provide an answer to the following questions:

1. What question were you trying to answer and why?

2. What did you do to answer your question and why?

3. What is your argument?

Your report should answer these questions in two pages or less. This report must be typed, and any diagrams, figures, or tables should be embedded into the document. Be sure to write in a persuasive style; you are trying to convince others that your claim is acceptable or valid!

Checkout Questions

Lab 8. Friction: Why Are Some Lubricants Better Than Others at Reducing the Coefficient of Friction Between Metal Plates?

1. Doubling the coefficient of friction between two surfaces will cut the frictional force in half.

   a. I agree with this statement.
   b. I disagree with this statement.

   Explain your answer, using an example from your investigation about friction and lubricants.

2. Most pickup trucks have four tires, but some have six. How do you think having two additional tires affects the frictional force between a truck and the road?

   a. Adding two tires increases the frictional force between the truck and the road.
   b. Adding two tires decreases the frictional force between the truck and the road.
   c. Adding two tires has no effect on the frictional force between the truck and the road.

   Explain your answer.

3. Societal needs and current events can influence the research that scientists and engineers decide to do.

   a. I agree with this statement.

   b. I disagree with this statement.

   Explain your answer, using an example from your investigation about friction and lubricants.

4. All scientists follow the same scientific method when doing research.

   a. I agree with this statement.

   b. I disagree with this statement.

   Explain your answer, using an example from your investigation about friction and lubricants.

5. Why is it useful to identify a system under study and then make a model of it during an investigation? In your answer, be sure to include examples from at least two different investigations.

6. Why is it important to think about the relationship between structure and function when trying to develop an explanation for a natural phenomenon? In your answer, be sure to include examples from at least two different investigations.

LAB 9

Lab 9. Falling Objects and Air Resistance: How Does the Surface Area of a Parachute Affect the Force Due to Air Resistance as an Object Falls Toward the Ground?

Introduction

When we solve motion problems in physics, we often neglect to take into account the effects of air resistance because, at slow speeds, they are relatively small compared with the force of gravity. Other times, we ignore air resistance when we perform calculations in order to simplify the problem. However, some devices like kites and parachutes are designed to use air resistance in order to function. In these cases, scientists need to account for the effect of air resistance on falling objects.

Besides being used for recreational purposes such as skydiving, parachutes play an important role in the humanitarian efforts of many governments. One of the first uses of parachutes to aid humanitarian efforts was the Berlin Airlift of 1948–1949 (*www.history.com/this-day-in-history/berlin-airlift-begins*). As tensions rose at the onset of the Cold War, the Soviet Union prevented any people or goods from entering West Berlin in Germany. In response, the United States and United Kingdom organized efforts to airdrop food, supplies, and coal (for power) into West Berlin. By the end of the Soviet blockade in 1949, over 200,000 flights had been made into and over Berlin.

An airdrop of food and medical supplies after a major earthquake in Haiti

The airdrop remains one of the more effective tools for bringing food and necessary supplies, such as medicine, to people that need it. Figure L9.1, for example, is a picture of the airdrop that took place in Haiti after the 2010 earthquake that nearly destroyed the city of Port-au-Prince.

Air resistance affects the net force acting on a falling object, although in some conditions the effect is negligible and/or not observable. Newton's second law states that the acceleration produced by a net force on an object is directly proportional to the magnitude of the net force, is in the same direction as the net force, and is inversely proportional to the mass of an object; or, in mathematical terms, acceleration equals net force divided by mass.

The acceleration of a falling object without air resistance is -9.8 m/s^2 because the net force acting on the falling object is equal to the force of gravity. However, when air resistance is present, then the net force on the object changes, because the force of air resistance counters the force of gravity.

An engineer needs to consider several different issues and work through a multistep design process in order to create a new parachute. The first step in the design process is to determine the performance specifications of the new parachute. This step requires the engineer to think about the minimum and maximum mass of any object that will be attached to the parachute and the maximum terminal velocity that the object will reach as it falls to the ground. Terminal velocity is the highest velocity attainable by an object as it falls through the air. Terminal velocity occurs when the drag force acting on the falling object is equal to the force of gravity. At this point, the sum of the forces acting on the object equals zero, and the resulting acceleration will be zero. A safe landing velocity for an object is usually between 2 and 5 m/s. The second step in the design process is to build a parachute with a specific surface area that will meet these important performance specifications. It is therefore important for engineers to understand how the surface area of a parachute affects the force of air resistance that acts on an object as it falls to the ground.

Your Task

Use what you know about forces and motion, structure and function, and models to design and carry out an investigation to determine how parachute surface area affects the force due to air resistance.

The guiding question of this investigation is, *How does the surface area of a parachute affect the force due to air resistance as an object falls toward the ground?*

Materials

You may use any of the following materials during your investigation:

Consumables	Equipment
• Large trash bags	• Safety glasses or goggles (required)
• Tape	• Electronic or triple beam balance
• String or fishing line	• Washers
	• Stopwatch
	• Ruler
	• Meterstick

If you have access to the following equipment, you may also consider using a video camera and a computer or tablet with video analysis software.

LAB 9

Safety Precautions

Follow all normal lab safety rules. In addition, take the following safety precautions:

1. Wear sanitized safety goggles or glasses during lab setup, hands-on activity, and takedown.

2. Do not throw the washers or the parachutes.

3. Do not stand on tables or chairs.

4. Wash hands with soap and water after completing the lab.

Investigation Proposal Required?   ☐ Yes   ☐ No

Getting Started

To answer the guiding question, you will need to design and carry out an experiment. To accomplish this task, you must determine what type of data you need to collect, how you will collect it, and how you will analyze it.

To determine *what type of data you need to collect*, think about the following questions:

- What are the boundaries and components of the system you are studying?
- How do the components of the system interact with each other?
- How might the structure of a parachute relate to its function?
- How will you determine the surface area of a parachute?
- How will you measure the force of air resistance?
- What will be the independent variable and the dependent variable for your experiment?

To determine *how you will collect the data*, think about the following questions:

- What conditions need to be satisfied to establish a cause-and-effect relationship?
- What measurement scale or scales should you use to collect data?
- What equipment will you need to make the measurements?
- What other variables will you need to control during your experiment?
- Do you need to include a control group?
- How will you make sure that your data are of high quality (i.e., how will you reduce error)?
- How will you keep track of and organize the data you collect?

To determine *how you will analyze the data*, think about the following questions:

- What type of calculations will you need to make?
- What types of models can you use to help you analyze the motion of a parachute?
- How could you use mathematics to describe a relationship between variables?
- What types of patterns might you look for as you analyze your data?
- Are there any proportional relationships that you can identify?
- What type of table or graph could you create to help make sense of your data?

Connections to the Nature of Scientific Knowledge and Scientific Inquiry

As you work through your investigation, you may want to consider

- how the culture of science, societal needs, and current events influence the work of scientists; and
- the role of imagination and creativity in science.

Initial Argument

Once your group has finished collecting and analyzing your data, your group will need to develop an initial argument. Your initial argument needs to include a claim, evidence to support your claim, and a justification of the evidence. The *claim* is your group's answer to the guiding question. The *evidence* is an analysis and interpretation of your data. Finally, the justification of the evidence is why your group thinks the evidence matters. The *justification* of the evidence is important because scientists can use different kinds of evidence to support their claims. Your group will create your initial argument on a whiteboard. Your whiteboard should include all the information shown in Figure L9.2.

FIGURE L9.2

Argument presentation on a whiteboard

The Guiding Question:	
Our Claim:	
Our Evidence:	Our Justification of the Evidence:

Argumentation Session

The argumentation session allows all of the groups to share their arguments. One or two members of each group will stay at the lab station to share that group's argument, while the other members of the group go to the other lab stations to listen to and critique the other arguments. This is similar to what scientists do when they propose, support, evaluate, and refine new ideas during a poster session at a conference. If you are presenting your group's argument, your goal is to share your ideas and answer questions. You should also keep a record of the critiques and suggestions made by your classmates so you can use this feedback to make your initial argument stronger. You can keep track of specific critiques and suggestions for improvement that your classmates mention in the space below.

LAB 9

Critiques about our initial argument and suggestions for improvement:

If you are critiquing your classmates' arguments, your goal is to look for mistakes in their arguments and offer suggestions for improvement so these mistakes can be fixed. You should look for ways to make your initial argument stronger by looking for things that the other groups did well. You can keep track of interesting ideas that you see and hear during the argumentation in the space below. You can also use this space to keep track of any questions that you will need to discuss with your team.

Interesting ideas from other groups or questions to take back to my group:

Once the argumentation session is complete, you will have a chance to meet with your group and revise your initial argument. Your group might need to gather more data or design a way to test one or more alternative claims as part of this process. Remember, your goal at this stage of the investigation is to develop the best argument possible.

Report

Once you have completed your research, you will need to prepare an *investigation report* that consists of three sections. Each section should provide an answer to the following questions:

1. What question were you trying to answer and why?

2. What did you do to answer your question and why?

3. What is your argument?

Your report should answer these questions in two pages or less. This report must be typed, and any diagrams, figures, or tables should be embedded into the document. Be sure to write in a persuasive style; you are trying to convince others that your claim is acceptable or valid!

Checkout Questions

Lab 9. Falling Objects and Air Resistance: How Does the Surface Area of a Parachute Affect the Force Due to Air Resistance as an Object Falls Toward the Ground?

1. Is there a maximum force due to air resistance that can act on a parachute?

   a. Yes
   b. No

   How do you know?

   What does your answer suggest about the effect of increasing the size of the parachute?

2. The equation for the force of air resistance (more formally, the drag) on a parachute is $F_D = C_D \rho v^2 A / 2$. In this equation, F_D is the drag force and v is the current velocity of the falling parachute and mass system. Is the drag force constant as a function of time?

   a. Yes
   b. No

   Justify your answer using the equation provided and/or data from your investigation.

3. Scientists share a set of values, norms, and commitments that shape what counts as knowing, how to represent or communicate information, and how to interact with other scientists.

    a. I agree with this statement.
    b. I disagree with this statement.

    Explain your answer, using an example from your investigation about air resistance and parachutes.

4. Scientists must use their imagination and creativity to figure out new ways to test ideas and collect or analyze data.

    a. I agree with this statement.
    b. I disagree with this statement.

    Explain your answer, using an example from your investigation about air resistance and parachutes.

5. Why is it useful to identify a system under study and then make a model of it during an investigation? In your answer, be sure to include examples from at least two different investigations.

6. Why is it important to think about the relationship between structure and function when trying to develop an explanation for a natural phenomenon? In your answer, be sure to include examples from at least two different investigations.

SECTION 4
Forces and Motion

Circular Motion and Rotation

Introduction Labs

LAB 10

Lab 10. Rotational Motion: How Do the Mass and the Distribution of Mass in an Object Affect Its Rotation?

Introduction

The wheel and axle are arguably among the most important inventions of all time. When a wheel turns around an axle, every point on the wheel moves in a circle about an axis of rotation. The motion around an axis of rotation is called rotational motion. We can describe the motion of a rotating object just like we can describe the motion of objects moving in a straight line. There are many similarities between rotational motion and linear motion. Rotating objects also have inertia and are influenced by the net force acting on them just like an object undergoing linear motion. A force that causes an object to rotate is called a torque.

There are also several important differences between rotational motion and linear motion. One big difference is how the distribution of mass within a rotating object affects how fast that object rotates around an axis of rotation. Take ice-skaters, gymnasts, and high divers as an example. These athletes often need to spin during a routine or a dive. As they spin, they can position their bodies in a way that will make them rotate faster or slower, which means they can change their angular velocity. Angular velocity is measured in radians per second (rad/s) and describes how much of a circle is completed by a rotating object in 1 second. A complete circle is equal to 360° or 2π radians (360° = 2π rad, therefore 1 radian ≈ 57.3°). An object with an angular velocity of 19 rad/s would complete approximately 3 rotations per second. Similar to linear motion, the mass of a rotating object influences how its velocity changes when a force acts on it. However, unlike linear motion, the angular velocity of a rotating object will also change when the position or distribution of the mass within that object is changed. Athletes, as a result, can change how far their body parts are away from the axis of rotation, which changes the distribution of their mass (their total mass stays constant, they do not gain or lose mass during the spin), in order to spin faster or slower during a routine or a dive (see Figure L10.1).

FIGURE L10.1

An ice-skater changes her body position to change the speed of her spin

Understanding rotational motion is important in science and has implications for aspects of our everyday lives. One implication related to rotational motion is the behavior of automobile wheels. The early inventors of the wheel were not concerned with traveling at the speeds we do today. Understanding how torque, mass, angular velocity, and angular acceleration influence the behavior of a wheel is important for ensuring that vehicles operate efficiently and are safe to drive. Putting too large a wheel on a car will cause the engine to become overworked and use more gas than necessary; additionally, if a wheel is too large or too massive, the brakes on the vehicle may not be strong enough to stop the wheel's rotation.

Your Task

Use what you know about rotational motion, patterns, and structure and function to design and conduct an investigation to determine how the mass and distribution of mass affect the rotation of an object.

The guiding question of this investigation is, ***How do mass and the distribution of mass in an object affect its rotation?***

Materials

You may use any of the following materials during your investigation:

Consumable	Equipment	
• Tape	• Safety glasses or goggles (required)	• Slotted masses, washers, or coins
	• Records	• Ramp (made from metersticks and a block)
	• Wood dowels	• Electronic or triple beam balance
	• Stopwatch	

If you have access to the following equipment, you may also consider using a video camera and a computer or tablet with video analysis software.

Safety Precautions

Follow all normal lab safety rules. In addition, take the following safety precautions:

1. Wear sanitized safety glasses or goggles during lab setup, hands-on activity, and takedown.

2. Keep fingers and toes out of the way of moving objects.

3. Wash hands with soap and water after completing the lab.

Investigation Proposal Required?   ☐ Yes   ☐ No

Getting Started

To answer the guiding question, you will need to design and carry out an experiment to determine how changes in mass and the distribution of mass affect the rotational motion of a record. The equipment setup shown in Figure L10.2 illustrates how you can use

LAB 10

FIGURE L10.2

Sample equipment setup using metersticks, a block, a wood dowel, and several records as shown from the side (a) and from the top (b)

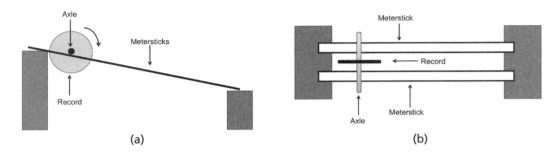

(a)   (b)

metersticks and blocks (or a stack of books) to build a ramp for a rolling record. You can use a wooden dowel rod that has the same diameter as the center hole of a record for the axle (add tape to the rod if it is too thin). You can then change the total mass of the rotating system by adding additional records to the axle or by adding masses to the record. You can tape washers or coins to the sides of the records at different spots on the record (e.g., close to the outer perimeter or close to the center axle) to change the distribution of mass in the rotating system. Before you begin designing your experiment using this equipment, however, you must first determine what type of data you need to collect, how you will collect it, and how you will analyze it.

To determine *what type of data you need to collect,* think about the following questions:

- What are the boundaries and components of the system you are studying?
- How do the components of the system interact with each other?
- How might changes to the structure of what you are studying change how it functions?
- How could you keep track of changes in this system quantitatively?
- What variables do you need to compare?
- What is the outcome variable for your investigation?

To determine *how you will collect the data,* think about the following questions:

- What measurement scale or scales should you use to collect data?
- What other variables will you need to control during your investigation?
- How will you make sure that your data are of high quality (i.e., how will you reduce error)?
- How will you keep track of and organize the data you collect?

To determine *how you will analyze the data*, think about the following questions:

- What types of patterns might you look for as you analyze your data?
- What type of calculations will you need to make?
- How could you use mathematics to describe a relationship between variables?
- What type of table or graph could you create to help make sense of your data?

Connections to the Nature of Scientific Knowledge and Scientific Inquiry

As you work through your investigation, you may want to consider

- the difference between observations and inferences in science, and
- the role of imagination and creativity in science.

Initial Argument

Once your group has finished collecting and analyzing your data, your group will need to develop an initial argument. Your argument must include a claim, evidence to support your claim, and a justification of the evidence. The *claim* is your group's answer to the guiding question. The *evidence* is an analysis and interpretation of your data. Finally, the *justification* of the evidence is why your group thinks the evidence matters. The justification of the evidence is important because scientists can use different kinds of evidence to support their claims. Your group will create your initial argument on a whiteboard. Your whiteboard should include all the information shown in Figure L10.3.

FIGURE L10.3 _____

Argument presentation on a whiteboard

The Guiding Question:	
Our Claim:	
Our Evidence:	Our Justification of the Evidence:

Argumentation Session

The argumentation session allows all of the groups to share their arguments. One or two members of each group will stay at the lab station to share that group's argument, while the other members of the group go to the other lab stations to listen to and critique the other arguments. This is similar to what scientists do when they propose, support, evaluate, and refine new ideas during a poster session at a conference. If you are presenting your group's argument, your goal is to share your ideas and answer questions. You should also keep a record of the critiques and suggestions made by your classmates so you can use this feedback to make your initial argument stronger. You can keep track of specific critiques and suggestions for improvement that your classmates mention in the space below.

Critiques about our initial argument and suggestions for improvement:

If you are critiquing your classmates' arguments, your goal is to look for mistakes in their arguments and offer suggestions for improvement so these mistakes can be fixed. You should look for ways to make your initial argument stronger by looking for things that the other groups did well. You can keep track of interesting ideas that you see and hear during the argumentation in the space below. You can also use this space to keep track of any questions that you will need to discuss with your team.

Interesting ideas from other groups or questions to take back to my group:

Once the argumentation session is complete, you will have a chance to meet with your group and revise your initial argument. Your group might need to gather more data or design a way to test one or more alternative claims as part of this process. Remember, your goal at this stage of the investigation is to develop the best argument possible.

Report

Once you have completed your research, you will need to prepare an investigation report that consists of three sections. Each section should provide an answer to the following questions:

1. What question were you trying to answer and why?

2. What did you do to answer your question and why?

3. What is your argument?

Your report should answer these questions in two pages or less. This report must be typed, and any diagrams, figures, or tables should be embedded into the document. Be sure to write in a persuasive style; you are trying to convince others that your claim is acceptable or valid!

LAB 10

Lab 10. Rotational Motion: How Do the Mass and the Distribution of Mass in an Object Affect Its Rotation?

Pictured below are four objects (a solid disc, a hoop, a sphere, and a solid cylinder). The mass and radius for each object are also provided. Use this information to answer question 1.

Solid disc	Hoop	Sphere	Solid cylinder
m = 350 g	m = 350 g	m = 350 g	m = 350 g
r = 15 cm	r = 15 cm	r = 15 cm	r = 15 cm
$I = \frac{1}{2}MR^2$	$I = MR^2$	$I = \frac{2}{5}MR^2$	$I = \frac{1}{2}MR^2$

1. If all four objects are released from the top of a ramp and allowed to roll to the bottom, what will be the finishing order for the objects?

   a. All objects will come to the bottom at the same time.

   b. The sphere will come to the bottom first, then the solid disc and solid cylinder at the same time, then the hoop.

   c. The hoop will come to the bottom first, then the solid disc and solid cylinder at the same time, then the sphere.

   d. The solid cylinder and the solid disc will come to the bottom first, then the sphere, then the hoop.

   Justify your answer using what you know about rotational motion.

2. Jeremy and Susan are playing on a merry-go-round. Susan says she wants to sit close to the center of the merry-go-round while Jeremy pushes the merry-go-round. Jeremy wants to sit closer to the outer edge. To get the merry-go-round to move at an angular velocity of one rotation per minute, who would need to turn the merry-go-round with a large force? Assume Jeremy and Susan each has a mass of 40 kg.

   a. Jeremy

   b. Susan

   c. The same force is needed.

   Use what you know about rotational motion to justify your answer.

3. Scientific research requires imagination and creativity.

   a. I agree with this statement.

   b. I disagree with this statement.

   Explain your answer, using an example from your investigation about the effect of a mass and the distribution of mass on a rotating object.

4. There is a difference between observations and inferences in science.

    a. I agree with this statement.

    b. I disagree with this statement.

Explain your answer, using an example from your investigation about the effect of a mass and the distribution of mass on a rotating object.

5. Why is it important to identify patterns and the underlying causes for those patterns in science? Be sure to include examples from at least two different investigations in your answer.

6. Why is useful to think about the relationship between structure and function during an investigation? Be sure to include examples from at least two different investigations in your answer.

Lab Handout

Lab 11. Circular Motion: How Does Changing the Angular Velocity of the Swinging Mass at the Top of a Whirligig and the Amount of Mass at the Bottom of a Whirligig Affect the Distance From the Top of the Tube to the Swinging Mass?

Introduction

Many things move in a circular path. For example, protons in the Large Hadron Collider, a car making a turn, and people on swing rides at amusement parks (see Figure L11.1) all move in a circular path. The orbital motion of many moons and planets is nearly circular as well. Even the stars in the spiral arms of the Milky Way galaxy, as they make their way around what may be a supermassive black hole at the galactic center, undergo circular motion.

Newton's first law of motion indicates that, without an unbalanced force acting on an object, an object moves in a straight line at constant speed. When we see an object traveling in a circular path, we must therefore assume that there is an unbalanced force acting on it. The unbalanced force is called a *centripetal* force. A force pulling toward the center of an object's circular path is what keeps it from moving in a straight line (*centripetal* comes from Latin, meaning "toward the center"). A centripetal force is whatever is causing the motion to be circular. For a planet, the centripetal force is gravity. For a turning car, friction between the tires and the road is the centripetal force. The string tension in the swing ride shown in Figure L11.1 is the force that keeps the people moving in a circle.

FIGURE L11.1 _____

The people on an a swing ride at an amusement park move in a circular path over time

We use the term *uniform circular motion* to describe the motion of an object when it follows the path of a circle with a constant speed. As it moves, it travels around the perimeter of the circle. The distance of one complete cycle around the perimeter of a circle is equal to the circumference of the circle. The average speed of an object in uniform circular motion can therefore be calculated using the following equation:

$$Average\ speed = \frac{2\pi\mathbf{r}}{T}$$

Velocity, unlike speed, has both a magnitude and a direction. The magnitude of the velocity vector is the instantaneous speed of the object. The velocity vector of an object in uniform circular motion is constant in magnitude but changing in direction. Since the object is constantly changing direction as it travels around the perimeter of a circle, it is also constantly accelerating. It is constantly accelerating because the direction of the velocity vector is constantly changing. This description of an object in uniform circular motion is also consistent with Newton's first and second laws. Because there is an unbalanced force acting on an object undergoing uniform circular motion, the object must also be accelerating.

In this investigation you will have an opportunity to learn more about the nature of uniform circular motion by attempting to explain how a whirligig works. A whirligig is a toy that has two masses attached to each other by a string. One mass, which is called the swinging mass, can be made to move in a stable circular path. Your goal is to develop a conceptual model of the whirligig that will allow you to explain what keeps the swinging mass moving in a stable circular path and predict the radius of the circle that it follows.

Your Task

Use what you know about uniform circular motion, centripetal force, vector and scalar quantities, and stability and change in systems to develop a conceptual model that will allow you to (a) explain what keeps the swinging mass at the end of a whirligig moving in a stable circular path and (b) predict the radius of the circle that the swinging mass follows when it is stable. To develop this conceptual model, you will also need to design and carry out an investigation to determine (a) the relationship between the radius of the circle that the swinging mass follows and its angular velocity and (b) the relationship between the radius of the circle that the swinging mass follows and the centripetal force caused by the mass at the bottom of the whirligig. You will then need to test it to determine if it is valid and acceptable.

The guiding question of this investigation is, *How does changing the angular velocity of the swinging mass at the top of a whirligig and the amount of mass at the bottom of a whirligig affect the distance from the top of the tube to the swinging mass?*

How Does Changing the Angular Velocity of the Swinging Mass at the Top of a Whirligig and the Amount of Mass at the Bottom of a Whirligig Affect the Distance From the Top of the Tube to the Swinging Mass?

Materials

You may use any of the following materials during your investigation:

- Safety glasses or goggles (required)
- Electronic or triple beam balance
- Hooked masses
- String or cord
- Rubber stopper
- PVC or metal tube
- Stopwatch

If you have access to the following equipment, you may also consider using a video camera and a computer or tablet with video analysis software.

Safety Precautions

Follow all normal lab safety rules. In addition, take the following safety precautions:

1. Wear sanitized safety glasses or goggles during lab setup, hands-on activity, and takedown.

2. Keep fingers and toes out of the way of moving objects.

3. Be aware of others around you when swinging the whirligig.

4. Wash hands with soap and water after completing the lab.

Investigation Proposal Required?   ☐ Yes   ☐ No

Getting Started

The first step in developing your conceptual model is to build a whirligig. You can make a whirligig using a rubber stopper, some string, a PVC or metal tube, and a hooked mass. Figure L11.2 shows how to assemble a whirligig with these materials. Once assembled, you can make the swinging mass (the rubber stopper in Figure L11.2) at the end of the whirligig move in a stable circle by changing the amount of mass you hang from the bottom of the whirligig or changing the angular velocity of the swinging mass.

You will then design and carry out two experiments using your whirligig. In the first experiment you will need to determine how changing the angular velocity of swinging mass affects the radius of the circle that the swinging mass follows when it is stable. You

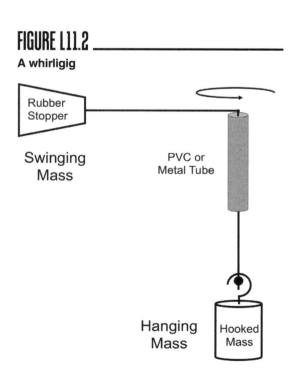

FIGURE L11.2 _____

A whirligig

Rubber Stopper

Swinging Mass

PVC or Metal Tube

Hanging Mass   Hooked Mass

will then need to conduct a second experiment to determine how changing the centripetal force caused by the mass at the bottom of the whirligig affects the radius of the circle that the swinging mass follows when it is stable. Before you can design these two experiments, however, you must determine what type of data you need to collect, how you will collect it, and how you will analyze it.

To determine *what type of data you need to collect,* think about the following questions:

- What are the boundaries and the components of the system under study?
- How can you describe the components of the system quantitatively?
- When is this phenomenon stable, and under which conditions does it change?
- Which factor(s) might control the rate of change in this system?
- How could you keep track of changes in this system quantitatively?
- How will you determine the radius of the stopper's orbit?
- How will you determine the stopper's velocity?
- What will be the independent variable and the dependent variable for each experiment?

To determine *how you will collect the data,* think about the following questions:

- Which variables are scalar quantities, and which variables are vector quantities?
- What other factors will you need to account for or control during each experiment?
- What measurement scale or scales should you use to collect data?
- What equipment will you need to collect the data?
- How will you make sure that your data are of high quality (i.e., how will you reduce error)?
- How will you keep track of and organize the data you collect?

To determine *how you will analyze the data,* think about the following questions:

- What types of patterns might you look for as you analyze your data?
- Are there any proportional relationships that you can identify?
- How could you use mathematics to describe a relationship between variables?
- What type of calculations will you need to make?
- What type of table or graph could you create to help make sense of your data?

Once you have determined (a) the relationship between the radius of the circle that the swinging mass follows and its angular velocity and (b) the relationship between the radius of the circle that the swinging mass follows and the centripetal force caused by the mass at

How Does Changing the Angular Velocity of the Swinging Mass at the Top of a Whirligig and the Amount of Mass at the Bottom of a Whirligig Affect the Distance From the Top of the Tube to the Swinging Mass?

the bottom of the whirligig, your group will need to develop your conceptual model. The model must be able to explain what keeps the swinging mass moving in a stable circular path and allow you to predict the radius of the circle that it follows. To be considered complete, your model must include information about the forces acting on the swinging mass. You should therefore consider including free-body diagrams in your model.

The last step in this investigation is to test your model. To accomplish this goal, you can add different hanging masses to the end of the whirligig (amounts that you did not test) to determine if your model of the motion of the swinging mass enables you to accurately predict the radius of the circle that the swinging mass follows when it is stable. If you are able to use your model to make accurate predictions about the radius of the circle that the swinging mass follows when it is stable, then you will be able to generate the evidence you need to convince others that your model is valid or acceptable.

Connections to the Nature of Scientific Knowledge and Scientific Inquiry

As you work through your investigation, you may want to consider

- the difference between laws and theories in science, and
- the difference between data and evidence.

Initial Argument

Once your group has finished collecting and analyzing your data, your group will need to develop an initial argument. Your argument must include a claim, evidence to support your claim, and a justification of the evidence. The *claim* is your group's answer to the guiding question. The *evidence* is an analysis and interpretation of your data. Finally, the *justification* of the evidence is why your group thinks the evidence matters. The justification of the evidence is important because scientists can use different kinds of evidence to support their claims. Your group will create your initial argument on a whiteboard. Your whiteboard should include all the information shown in Figure L11.3.

FIGURE L11.3 _____

Argument presentation on a whiteboard

The Guiding Question:	
Our Claim:	
Our Evidence:	Our Justification of the Evidence:

Argumentation Session

The argumentation session allows all of the groups to share their arguments. One or two members of each group will stay at the lab station to share that group's argument, while the other members of the group go to the other lab stations to listen to and critique the

LAB 11

other arguments. This is similar to what scientists do when they propose, support, evaluate, and refine new ideas during a poster session at a conference. If you are presenting your group's argument, your goal is to share your ideas and answer questions. You should also keep a record of the critiques and suggestions made by your classmates so you can use this feedback to make your initial argument stronger. You can keep track of specific critiques and suggestions for improvement that your classmates mention in the space below.

Critiques about our initial argument and suggestions for improvement:

If you are critiquing your classmates' arguments, your goal is to look for mistakes in their arguments and offer suggestions for improvement so these mistakes can be fixed. You should look for ways to make your initial argument stronger by looking for things that the other groups did well. You can keep track of interesting ideas that you see and hear during the argumentation in the space below. You can also use this space to keep track of any questions that you will need to discuss with your team.

Interesting ideas from other groups or questions to take back to my group:

Circular Motion

How Does Changing the Angular Velocity of the Swinging Mass at the Top of a Whirligig and the Amount of Mass at the Bottom of a Whirligig Affect the Distance From the Top of the Tube to the Swinging Mass?

Once the argumentation session is complete, you will have a chance to meet with your group and revise your initial argument. Your group might need to gather more data or design a way to test one or more alternative claims as part of this process. Remember, your goal at this stage of the investigation is to develop the best argument possible.

Report

Once you have completed your research, you will need to prepare an investigation report that consists of three sections. Each section should provide an answer to the following questions:

1. What question were you trying to answer and why?

2. What did you do to answer your question and why?

3. What is your argument?

Your report should answer these questions in two pages or less. This report must be typed, and any diagrams, figures, or tables should be embedded into the document. Be sure to write in a persuasive style; you are trying to convince others that your claim is acceptable or valid!

LAB 11

Lab 11. Circular Motion: How Does Changing the Angular Velocity of the Swinging Mass at the Top of a Whirligig and the Amount of Mass at the Bottom of a Whirligig Affect the Distance From the Top of the Tube to the Swinging Mass?

1. Sketch a free-body diagram for a stopper swung around on a whirligig.

2. Sketch a free-body diagram for the hanging mass on a whirligig.

3. Sketch a free-body diagram for a car on a flat road making a turn.

Circular Motion

How Does Changing the Angular Velocity of the Swinging Mass at the Top of a Whirligig and the Amount of Mass at the Bottom of a Whirligig Affect the Distance From the Top of the Tube to the Swinging Mass?

4. Sketch a free-body diagram for Jupiter orbiting the Sun.

5. How would a whirligig need to be swung to support a heavier hanging mass?

    a. Faster
    b. Slower

    Explain your reasoning.

6. Theories explain a phenomenon and laws describe the behavior of a phenomenon.

    a. I agree with this statement.
    b. I disagree with this statement.

    Explain your answer, using an example from your investigation about the whirligig.

7. Evidence is data collected during an investigation.

    a. I agree with this statement.
    b. I disagree with this statement.

    Explain your answer, using an example from your investigation about the whirligig.

8. Why is important to identify factors that contribute to the stability of a system? Be sure to include examples from at least two different investigations in your answer.

9. Why is it useful to create models of a system during an investigation? Be sure to include examples from at least two different investigations in your answer.

Application Lab

LAB 12

Lab 12. Torque and Rotation: How Can Someone Predict the Amount of Force Needed to Open a Bottle Cap?

Introduction

With the invention of the modern twist cap, there have been major advances in the ways that food and beverages can be packaged; in fact, it is difficult to find a bottle that does not close by using a twist cap. Common twist caps are used to seal soda bottles, gallon milk jugs, jars of pasta sauce, and even small bottles of medicine. Twist caps lock onto bottles and jars using threads—like a screw—on the underside of the cap; the outside of the bottle opening also has a set of threads. When the cap is placed over the bottle opening, the threads match up, and twisting the cap causes the threads to lock together and form a tight seal on the bottle. The threads of the twist cap and the threads on the bottle fit together like two ramps sliding past each other. The friction between the thread surfaces keeps the cap from coming loose by accident. Figure L12.1 shows how the threads of the cap and bottle fit together.

FIGURE L12.1

Cross-section of a threaded twist cap being screwed onto a threaded bottle

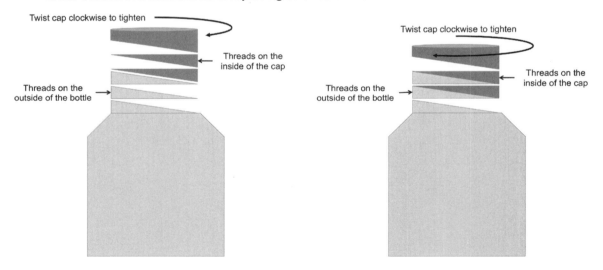

When companies are packaging different substances in bottles and jars that use this twist-top style of lid, they must ensure that the lids are installed with the proper amount of torque so that they form a good seal. A proper seal ensures that nothing leaks out and the food or drink stays fresh. When a net torque is applied to an object, the object will rotate.

The rotation of the object is related to two primary factors: (a) how strong of a force was applied to the object and (b) the location where the force was applied.

Think about closing the door to your science classroom. If you close the door by pushing on the handle, the door closes very easily. In this example, a small force is required when it is applied far away from the axis of rotation (the hinges of the door). When the same force is applied to the door, but halfway between the hinges and the handle, the door may not close. Because the location of the applied force is closer to the axis of rotation, the applied torque is decreased and the door may not close. If you want to close the door by applying a force at a location closer to the hinges, you likely need to apply a greater force.

Torque, and the relationship between the location of the applied force and the magnitude of the applied force, is important across many other relationships. Along with opening and closing bottle caps, torque plays an important role in tightening the lug nuts on your car's wheels, turning on a water faucet, and turning door handles. Part of the design process of tools and devices that twist is finding a balance between the location where a person will apply a force and the force needed to create a large enough torque. If you were to take apart a door handle, the part that actually turns to open your door is a small rod. The bulk of the handle is there to change the amount of force required to open the door.

Your Task

Use what you know about torque, proportional relationships, and systems and system models to design an investigation to determine the relationship between the force applied to open a bottle cap and the location from the axis of rotation the force is applied. Your goal is to create a mathematical model that will allow you to predict where a force must be applied on a bottle cap to open the cap for any specified force.

The guiding question of this investigation is, ***How can someone predict the amount of force needed to open a bottle cap?***

Materials

You may use any of the following materials during your investigation:

Consumables	Equipment
• String	• Safety glasses or goggles (required)
• Tape	• Plastic bottle modified with ruler or meterstick attached to the cap
	• Spring scales (or force sensor with interface)
	• Meterstick
	• Hanging mass set
	• Table clamp

LAB 12

Safety Precautions

Follow all normal lab safety rules. In addition, take the following safety precautions:

1. Wear sanitized safety glasses or goggles during lab setup, hands-on activity, and takedown.

2. Wash hands with soap and water after completing the lab.

Investigation Proposal Required?   ☐ Yes   ☐ No

Getting Started

To answer the guiding question, you will need to design and carry out an experiment. To accomplish this task, you must determine what type of data you need to collect, how you will collect it, and how you will analyze it.

To determine *what type of data you need to collect*, think about the following questions:

- What are the boundaries and components of the system you are studying?
- How do the components of the system interact with each other?
- How could you keep track of changes in this system quantitatively?
- How will you know the torque required to unscrew the cap?
- What variables do you need to measure to calculate torque?

To determine *how you will collect the data*, think about the following questions:

- What scale or scales should you use when you take your measurements?
- What equipment will you need to collect the data?
- How will you make sure that your data are of high quality (i.e., how will you reduce error)?
- How will you keep track of the data you collect?
- How will you organize your data?

To determine *how you will analyze the data*, think about the following questions:

- What types of patterns might you look for as you analyze your data?
- Are there any proportional relationships that you can identify?
- What type of calculations will you need to make?
- What type of table or graph could you create to help make sense of your data?
- What types of mathematical relationships might you use to model the system under study?

Connections to the Nature of Scientific Knowledge and Scientific Inquiry

As you work through your investigation, you may want to consider

- the difference between data and evidence in science, and
- the role of imagination and creativity in science.

Initial Argument

Once your group has finished collecting and analyzing your data, your group will need to develop an initial argument. Your argument must include a claim, evidence to support your claim, and a justification of the evidence. The *claim* is your group's answer to the guiding question. The *evidence* is an analysis and interpretation of your data. Finally, the *justification* of the evidence is why your group thinks the evidence matters. The justification of the evidence is important because scientists can use different kinds of evidence to support their claims. Your group will create your initial argument on a whiteboard. Your whiteboard should include all the information shown in Figure L12.2.

FIGURE L12.2

Argument presentation on a whiteboard

The Guiding Question:	
Our Claim:	
Our Evidence:	Our Justification of the Evidence:

Argumentation Session

The argumentation session allows all of the groups to share their arguments. One or two members of each group will stay at the lab station to share that group's argument, while the other members of the group go to the other lab stations to listen to and critique the other arguments. This is similar to what scientists do when they propose, support, evaluate, and refine new ideas during a poster session at a conference. If you are presenting your group's argument, your goal is to share your ideas and answer questions. You should also keep a record of the critiques and suggestions made by your classmates so you can use this feedback to make your initial argument stronger. You can keep track of specific critiques and suggestions for improvement that your classmates mention in the space below.

Critiques about our initial argument and suggestions for improvement:

LAB 12

If you are critiquing your classmates' arguments, your goal is to look for mistakes in their arguments and offer suggestions for improvement so these mistakes can be fixed. You should look for ways to make your initial argument stronger by looking for things that the other groups did well. You can keep track of interesting ideas that you see and hear during the argumentation in the space below. You can also use this space to keep track of any questions that you will need to discuss with your team.

Interesting ideas from other groups or questions to take back to my group:

Once the argumentation session is complete, you will have a chance to meet with your group and revise your initial argument. Your group might need to gather more data or design a way to test one or more alternative claims as part of this process. Remember, your goal at this stage of the investigation is to develop the best argument possible.

Report

Once you have completed your research, you will need to prepare an investigation report that consists of three sections. Each section should provide an answer to the following questions:

1. What question were you trying to answer and why?

2. What did you do to answer your question and why?

3. What is your argument?

Your report should answer these questions in two pages or less. This report must be typed, and any diagrams, figures, or tables should be embedded into the document. Be sure to write in a persuasive style; you are trying to convince others that your claim is acceptable or valid!

Checkout Questions

Lab 12. Torque and Rotation: How Can Someone Predict the Amount of Force Needed to Open a Bottle Cap?

1. A contestant on a game show was spinning a large wheel to try and win money. The contestant was spinning the wheel as hard as he could, but the wheel only spun around a couple of times. The contestant suggested moving the handles closer to the center so that he and other players could make the wheel spin more with each push. The suggested change is shown below.

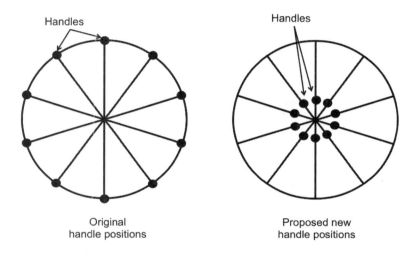

Original
handle positions

Proposed new
handle positions

Will moving the handles closer to the center of the wheel help the contestants get more spins per push?

   a. Yes

   b. No

Explain your answer, using what you know about torque and rotational motion.

2. The bones and muscles found in humans and chimpanzees are almost identical. Chimpanzees, however, are much stronger than humans even though they are smaller. One explanation for this difference in strength is that the muscles of

chimpanzees are attached to the bones in slightly different ways than those of humans; therefore, chimpanzees are able to generate a greater torque with smaller muscles. The diagram below shows where the bicep muscle is attached to the upper and lower limb in a human arm and in a chimpanzee arm. When thinking about this, remember that muscles can only contract, so a bicep muscle pulls the lower limb toward the upper limb.

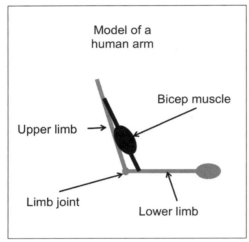
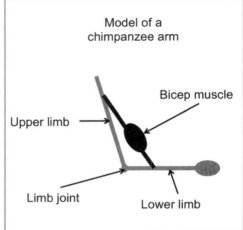

Use what you know about torque to explain why the difference could result in greater torque.

3. In science, it is possible for a variable to be proportionally related to two other variables.

   a. I agree with this statement.

   b. I disagree with this statement.

Explain your answer, using an example from your investigation about torque and the rotation of bottle caps.

4. Science requires imagination and creativity.

    a. I agree with this statement.

    b. I disagree with this statement.

Explain your answer, using an example from your investigation about torque and the rotation of bottle caps.

5. There is a difference between data and evidence in science. Explain what data and evidence are and how they are different from each other, using an example from your investigation about torque and the rotation of bottle caps.

6. In science, identifying the system under study is a prerequisite for being able to mathematically model the system. Explain why this statement is true, using an example from your investigation about torque and the rotation of bottle caps.

SECTION 5
Forces and Motion

Oscillations

Introduction Labs

Lab 13. Simple Harmonic Motion and Pendulums: What Variables Affect the Period of a Pendulum?

Introduction

A pendulum, which is a mass swinging at the end of a rope, has a wide range of uses in our daily lives. One of the most frequent uses of a pendulum is in a clock. Christiaan Huygens built the first pendulum clock in 1656 (see Figure L13.1 and version available at *www.nsta.org/adi-physics1*), and his use of a pendulum to keep accurate time was considered a breakthrough in clock design. Pendulums can also be found as parts of amusement park rides, in religious ceremonies, and in tools that help musicians keep a beat. Most school-age children are also familiar with pendulums, because playground swings are just a pendulum with a person at one end.

Pendulums are part of a class of objects that undergo simple harmonic motion; such objects are called oscillators. Harmonic oscillators are objects that move about a point called the equilibrium position (see Figure L13.2). When a pendulum is not moving, the bob will rest (or hang motionless) at the equilibrium position. When an outside force moves the bob from its equilibrium position, a restoring force causes the object to move back toward its equilibrium position. This process is then repeated multiple times as the bob swings back and forth. This motion is referred to as *simple,* because after the initial force to move the bob from equilibrium, the only forces acting on the bob are the restoring force and the tension in the string. Other types of harmonic motion are called *damped,* when friction slows down the motion, or *driven,* when an outside force is repeatedly exerted on the oscillator. There

FIGURE L13.1 _____

The original pendulum clock built by Christiaan Huygens in 1656

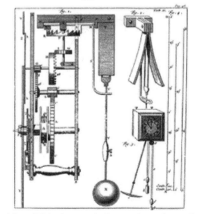

Note: This image is best viewed on the book's Extras page at *www.nsta.org/ adi-physics1*.

FIGURE L13.2 _____

The components of a pendulum

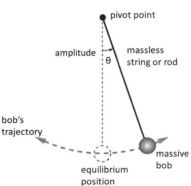

are many ways to describe the motion of a bob. The most frequent is the period (*T*), which is how long it takes a bob to make one full swing back and forth.

For most pendulums, the period does not change from one swing to the next. This makes the pendulum a particularly useful tool for timekeeping, such as in the pendulum clock shown in Figure L13.1. Early physicists recognized this and investigated the pendulum to understand what variables influence its period. This would allow them to more effectively use pendulums in clocks, as well as in other devices.

Your Task

Use what you know about simple harmonic motion, causal relationships, the relationship between structure and function in nature, and the importance of patterns to design and carry out a series of experiments to determine which variables do and which variables do not change the period of the pendulum.

The guiding question of this investigation is, **What variables affect the period of a pendulum?**

Materials

You may use any of the following materials during your investigation:

Consumables
- Tape
- String

Equipment
- Safety glasses or goggles (required)
- Electronic or triple beam balance
- Washers
- Paper clips
- Protractor
- Ruler
- Meterstick
- Stopwatch
- Scissors

To use a photogate system, you will need to have a sensor interface and a computer, tablet, or graphing calculator with data collection and analysis software. To use video analysis, you will need to have a video camera and a computer or tablet with video analysis software.

Safety Precautions

Follow all normal lab safety rules. In addition, take the following safety precautions:

1. Wear sanitized safety glasses or goggles during lab setup, hands-on activity, and takedown.

2. Keep fingers and toes out of the way of moving objects.

3. Use caution when working with scissors. They are sharp and can cut or puncture skin.

LAB 13

4. Wash hands with soap and water after completing the lab.

Investigation Proposal Required?   ☐ Yes   ☐ No

Getting Started

To answer the guiding question, you will need to design and carry out several different experiments. Each experiment should look at one potential variable that may or may not affect the period of a pendulum. Some potential variables include the mass of the bob, the length of the pendulum, and the release angle. For each of your experiments, you must determine what type of data you need to collect, how you will collect it, and how you will analyze it before you begin.

To determine *what type of data you need to collect*, think about the following questions:

- What are the boundaries and components of the system you are studying?
- How can you describe the components of the system quantitatively?
- How could you keep track of changes in this system quantitatively?
- How might changes to the structure of pendulum change how it functions?
- What might be the underlying cause of a change in the period of a pendulum?
- What will be the independent variable and the dependent variable for each experiment?

To determine *how you will collect the data*, think about the following questions:

- How will you set up your pendulum?
- How will you measure the period of the pendulum?
- What will you need to hold constant during each experiment?
- What conditions need to be satisfied to establish a cause-and-effect relationship?
- What measurement scale or scales should you use to collect data?
- How will you make sure that your data are of high quality (i.e., how will you reduce error)?
- How will you keep track of and organize the data you collect?

To determine *how you will analyze the data*, think about the following questions:

- What type of calculations will you need to make?
- What types of patterns might you look for as you analyze your data?
- How could you use mathematics to show a cause-and-effect relationship?
- What type of table or graph could you create to help make sense of your data?

Connections to the Nature of Scientific Knowledge and Scientific Inquiry

As you work through your investigation, you may want to consider

- the difference between data and evidence in science, and
- the nature and role of experiments in science.

Initial Argument

Once your group has finished collecting and analyzing your data, your group will need to develop an initial argument. Your initial argument needs to include a claim, evidence to support your claim, and a justification of the evidence. The *claim* is your group's answer the guiding question. The *evidence* is an analysis and interpretation of your data. Finally, the *justification* of the evidence is why your group thinks the evidence matters. The justification of the evidence is important because scientists can use different kinds of evidence to support their claims. Your group will create your initial argument on a whiteboard. Your whiteboard should include all the information shown in Figure L13.3.

FIGURE L13.3 _____
Argument presentation on a whiteboard

The Guiding Question:	
Our Claim:	
Our Evidence:	Our Justification of the Evidence:

Argumentation Session

The argumentation session allows all of the groups to share their arguments. One or two members of each group will stay at the lab station to share that group's argument, while the other members of the group go to the other lab stations to listen to and critique the other arguments. This is similar to what scientists do when they propose, support, evaluate, and refine new ideas during a poster session at a conference. If you are presenting your group's argument, your goal is to share your ideas and answer questions. You should also keep a record of the critiques and suggestions made by your classmates so you can use this feedback to make your initial argument stronger. You can keep track of specific critiques and suggestions for improvement that your classmates mention in the space below.

Critiques about our initial argument and suggestions for improvement:

If you are critiquing your classmates' arguments, your goal is to look for mistakes in their arguments and offer suggestions for improvement so these mistakes can be fixed. You should look for ways to make your initial argument stronger by looking for things that the other groups did well. You can keep track of interesting ideas that you see and hear during the argumentation in the space below. You can also use this space to keep track of any questions that you will need to discuss with your team.

Interesting ideas from other groups or questions to take back to my group:

Once the argumentation session is complete, you will have a chance to meet with your group and revise your initial argument. Your group might need to gather more data or design a way to test one or more alternative claims as part of this process. Remember, your goal at this stage of the investigation is to develop the best argument possible.

Report

Once you have completed your research, you will need to prepare an *investigation report* that consists of three sections. Each section should provide an answer to the following questions:

1. What question were you trying to answer and why?

2. What did you do to answer your question and why?

3. What is your argument?

Your report should answer these questions in two pages or less. This report must be typed, and any diagrams, figures, or tables should be embedded into the document. Be sure to write in a persuasive style; you are trying to convince others that your claim is acceptable or valid!

Reference

Jeremy Norman's HistoryofInformation.com. 2016. Huygens invents the pendulum clock, increasing accuracy sixty fold (1656). *www.historyofinformation.com/expanded.php?id=3506.*

Lab 13. Simple Harmonic Motion and Pendulums: What Variables Affect the Period of a Pendulum?

1. The equation for the period of a pendulum is $T = 2\pi\sqrt{(L/\mathbf{g})}$, where T is the period, L is the length of the pendulum, and $\mathbf{g}$ is the acceleration due to gravity. If a person were to take a pendulum to the Moon, which has a gravitation pull approximately one-sixth that of Earth, what would happen to the period of the pendulum?

   a. The period would increase.

   b. The period would decrease.

   c. The period would stay the same.

   How do you know?

2. Why does the mass of bob have no effect on the period of a pendulum?

LAB 13

3. It is equally important for scientists to identify variables that do have a cause-and-effect relationship and those variables that do not have a cause-and-effect relationship.

    a. I agree with this statement.
    b. I disagree with this statement.

    Explain your answer, using an example from your investigation about pendulums.

4. Scientists use the term *data* when they are talking about observations and the term *evidence* when they are talking about measurements.

    a. I agree with this statement.
    b. I disagree with this statement.

    Explain your answer, using an example from your investigation about pendulums.

5. Why is important to look for patterns in science? In your answer, be sure to include one example from your investigation on pendulums and at least one more example from another investigation you have conducted in either this class or another science class.

6. Scientists often examine the structure of an object or material during an investigation. Explain why it is useful to examine the structure of an object or material, using an example from your investigation about pendulums.

7. Experiments are one of the most powerful approaches to answering questions in science. Identify the components of an experiment and explain why they are so important in science, using an example from your investigation about pendulums.

LAB 14

Lab Handout

Lab 14. Simple Harmonic Motion and Springs: What Is the Mathematical Model of the Simple Harmonic Motion of a Mass Hanging From a Spring?

Introduction

A basic but important kind of motion is called simple harmonic motion. Simple harmonic motion is a type of periodic or oscillatory motion. An example of simple harmonic motion is the way a mass moves up and down when it is attached to a spring (see Figure L14.1). When the mass attached to a spring is not moving, the mass is said to be at equilibrium. When the mass is moved from the equilibrium position (such as when a person pulls down on the spring), a force from the spring (F_k) acts on the mass to restore it to the equilibrium position. This force is called the restoring force. The further the mass is moved from the equilibrium position, the greater the restoring force becomes. In fact, the magnitude of the restoring force is directly proportional to the distance the mass is moved away from the equilibrium position (the displacement) and acts in the direction opposite to the direction of displacement.

We can explain the underlying cause of simple harmonic motion using Newton's second law of motion. Newton's second law of motion indicates that an object will accelerate when acted on by an unbalanced force. Thus, when an oscillator (in this case a mass attached to a spring) is disturbed from equilibrium, it accelerates in the general direction of the equilibrium position. When the mass reaches the equilibrium position, it has a non-zero velocity and the sum of the forces acting on it is zero. The mass will therefore move through the equilibrium position, at which point the restoring force increases until the mass reaches a maximum displacement from the equilibrium position. At the maximum displacement position, the restoring force is also at a maximum, while the velocity is equal to zero (0 m/s). This process repeats over time, which results in periodic or oscillatory motion.

In this investigation you will have an opportunity to examine the simple harmonic motion of a mass hanging on a spring. Your goal is to create a mathematical model that you can use to describe the vertical position of the mass in terms of time. Therefore, you will need to investigate the effect of different masses, release points, and types of springs during your investigation. It is important to note that your model will ignore dampening, the effect of slowing the mass-spring system down to a stop by frictional forces.

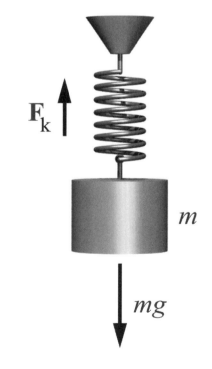

FIGURE L14.1
A mass on a spring

F_k

m

mg

Your Task

Use what you know about simple harmonic motion, patterns, and stability and change in systems to develop a function that will allow you to model the motion of a mass hanging from a spring. To develop a mathematical model, you will need to design and carry out several experiments to determine how (a) different masses, (b) different release points, and (c) different spring types affect the motion of a mass hanging from a spring. Once you have developed your model, you will need to test it to determine if allows you to make accurate predictions about the vertical position of the mass in terms of time.

The guiding question of this investigation is, *What is the mathematical model of the simple harmonic motion of a mass hanging from a spring?*

Materials

You may use any of the following materials during your investigation:

- Safety glasses or goggles (required)
- Support stand
- Suspension hook clamp
- Hanging mass set
- Springs (variety)

- Motion detector/sensor
- Interface for motion detector/sensor
- Computer, tablet, or graphing calculator with data collection and analysis software

Safety Precautions

Follow all normal lab safety rules. In addition, take the following safety precautions:

1. Wear sanitized safety glasses or goggles during lab setup, hands-on activity, and takedown.

2. Keep fingers and toes out of the way of moving objects.

3. Wash hands with soap and water after completing the lab.

Investigation Proposal Required?   ☐ Yes   ☐ No

Getting Started

The first step in developing your mathematical model is to design and carry out three experiments. In the first experiment, you will need to determine how changing the mass affects the motion of a mass-spring system. You will then need to determine how changing the release point of the mass affects the motion of the mass-spring system. Finally, you will need to determine how changing the type of spring affects the motion of the mass-spring system. Figure L14.2 illustrates how you can use the available equipment to study the motion of a mass-spring system in each experiment. Before you can design your experiments, however, you must determine what type of data you need to collect, how you will collect it, and how you will analyze it.

To determine *what type of data you need to collect,* think about the following questions:

- What are the boundaries and components of the mass-spring system you are studying?
- Which factor(s) might control the rate of change in the mass-spring system?
- How could you keep track of changes in this system quantitatively?
- Under what conditions is the system stable, and under what conditions does it change?
- How will you measure the vertical position of the mass over time?
- What will be the independent variables and the dependent variables for each experiment?

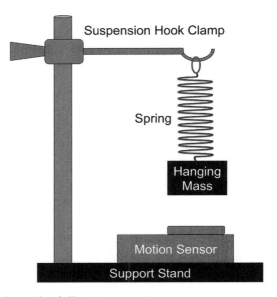

One way to examine the motion of the mass-spring system using the available equipment

Suspension Hook Clamp

Spring

Hanging Mass

Motion Sensor

Support Stand

To determine *how you will collect the data,* think about the following questions:

- What other factors will you need to control during each experiment?
- What scale or scales should you use when you take your measurements?
- How will you make sure that your data are of high quality (i.e., how will you reduce error)?
- How will you keep track of and organize the data you collect?

To determine *how you will analyze the data,* think about the following questions:

- How much of your data is useful, given that you want to ignore dampening?
- What type of calculations will you need to make?
- What types of patterns might you look for as you analyze your data?
- What type of table or graph could you create to help make sense of your data?
- What types of equations can you use to describe motion that is periodic or harmonic?

Once you have determined how different masses, release points, and spring types affect the motion of a mass-spring system, your group will need to develop a mathematical model. The model must allow you to make accurate predictions about the vertical position of the mass in terms of time.

The last step in this investigation will be to test your model. To accomplish this goal, you can add different hanging masses (amounts that you did not test) to the end of one of the springs or try different release points (ones that you did not test) to determine whether your mathematical model helps you make accurate predictions. If you are able to use your model to make accurate predictions, then you will be able to generate the evidence you need to convince others that it is a valid and acceptable model of the simple harmonic motion of a mass hanging from a spring.

Connections to the Nature of Scientific Knowledge and Scientific Inquiry

As you work through your investigation, you may want to consider

- how scientists use different methods to answer different types of questions, and
- the role of imagination and creativity in science.

Initial Argument

Once your group has finished collecting and analyzing your data, your group will need to develop an initial argument. Your argument must include a claim, evidence to support your claim, and a justification of the evidence. The *claim* is your group's answer to the guiding question. The *evidence* is an analysis and interpretation of your data. Finally, the *justification* of the evidence is why your group thinks the evidence matters. The justification of the evidence is important because scientists can use different kinds of evidence to support their claims. Your group will create your initial argument on a whiteboard. Your whiteboard should include all the information shown in Figure L14.3.

FIGURE L14.3

Argument presentation on a whiteboard

The Guiding Question:	
Our Claim:	
Our Evidence:	Our Justification of the Evidence:

Argumentation Session

The argumentation session allows all of the groups to share their arguments. One or two members of each group will stay at the lab station to share that group's argument, while the other members of the group go to the other lab stations to listen to and critique the other arguments. This is similar to what scientists do when they propose, support, evaluate, and refine new ideas during a poster session at a conference. If you are presenting your group's argument, your goal is to share your ideas and answer questions. You should also keep a record of the critiques and suggestions made by your classmates so you can use this feedback to make your initial argument stronger. You can keep track of specific critiques and suggestions for improvement that your classmates mention in the space below.

LAB 14

Critiques about our initial argument and suggestions for improvement:

If you are critiquing your classmates' arguments, your goal is to look for mistakes in their arguments and offer suggestions for improvement so these mistakes can be fixed. You should look for ways you to make your initial argument stronger by looking for things that the other groups did well. You can keep track of interesting ideas that you see and hear during the argumentation in the space below. You can also use this space to keep track of any questions that you will need to discuss with your team.

Interesting ideas from other groups or questions to take back to my group:

Once the argumentation session is complete, you will have a chance to meet with your group and revise your initial argument. Your group might need to gather more data or design a way to test one or more alternative claims as part of this process. Remember, your goal at this stage of the investigation is to develop the best argument possible.

Report

Once you have completed your research, you will need to prepare an *investigation report* that consists of three sections. Each section should provide an answer to the following questions:

1. What question were you trying to answer and why?

2. What did you do to answer your question and why?

3. What is your argument?

Your report should answer these questions in two pages or less. This report must be typed, and any diagrams, figures, or tables should be embedded into the document. Be sure to write in a persuasive style; you are trying to convince others that your claim is acceptable or valid!

LAB 14

Lab 14. Simple Harmonic Motion and Springs: What Is the Mathematical Model of the Simple Harmonic Motion of a Mass Hanging From a Spring?

In most physics textbooks, the position of an object in simple harmonic motion is described using the equation below:

$$\mathbf{x} = A\cos(\omega t + \delta)$$

1. Given your model in terms of A, B, C, and D, define the parameters A, ω, and δ and discuss the meaning of each in terms of the position of the mass on a spring.

Use the graph below to answer questions 2 and 3.

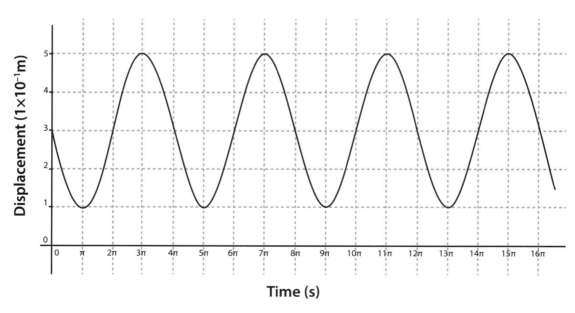

2. What are the parameters A, B, C, and D for the general model $\mathbf{s}(t) = \mathbf{C} + A\cos B\,(t - D)$?

3. What are the parameters for A, ω, and δ for the model $x = A\cos(\omega t + \delta)$?

4. Scientists do not need to be creative or have a good imagination.

    a. I agree with this statement.
    b. I disagree with this statement.

    Explain your answer, using an example from your investigation about simple harmonic motion and springs.

5. It is important to understand what makes a system stable or unstable and what contributes to the rates of change in a system.

   a. I agree with this statement.
   b. I disagree with this statement.

   Explain your answer, using an example from your investigation about simple harmonic motion and springs.

6. Scientists use different methods to answer different types of questions. Explain how the type of question a scientist asks affects the methods he or she uses to answer those questions, using an example from your investigation about simple harmonic motion and springs.

7. Scientists often look for and try to explain patterns in nature. Explain why it is useful to look for and explain patterns in nature, using an example from your investigation about simple harmonic motion and springs.

8. Models in science can be physical, conceptual, or mathematical. Explain the difference in these types of models and discuss the strengths and weaknesses of each type of model, using an example from your investigation about simple harmonic motion and springs.

Application Lab

LAB 15

Lab 15. Simple Harmonic Motion and Rubber Bands: Under What Conditions Do Rubber Bands Obey Hooke's Law?

Introduction

Harmonic oscillators are objects that move about an equilibrium point due to a restoring force. Two of the most common harmonic oscillators are pendulums and springs. Many objects that we encounter on a daily basis either contain a pendulum, such as a grandfather clock, or are a pendulum, such as the swings found on playgrounds. There are also many other objects that we use that incorporate springs. The shock absorbers found on cars, for example, are just large springs. Pendulums and springs, however, are not the only harmonic oscillators that are found in the world around us. For example, tire swings are a type of harmonic oscillator called a torsional oscillator because the restoring force from the rope causes the tire swing to twist back and forth (torsional motion is the motion of twisting).

Another object that oscillates is a bungee jump ride (see Figure L15.1). When a person makes a bungee jump, he or she is attached to bungee cord and then dropped from a considerable height. That person then bounces up and down for the duration of the ride.

Many oscillators obey the equations that describe simple harmonic motion. One of those equations is called Hooke's law, which states that the farther the oscillator is moved from its equilibrium point, the greater the restoring force on the oscillator. More specifically, the restoring force is directly proportional to the displacement the oscillator is from equilibrium. The equation for Hooke's law is $F = -k\mathbf{x}$, where $\mathbf{F}$ is the restoring force, $\mathbf{x}$ is the displacement from equilibrium, and k is a constant of proportionality. This equation includes a negative sign because the restoring force always acts in the direction opposite the direction of displacement. For example, if the bungee cord is pulled in the down direction, then the restoring force acts in the up direction.

FIGURE L15.1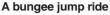
A bungee jump ride

It is important to note, however, that Hooke's law is not valid for all ranges of possible displacements. For example, if a spring is pulled too far from its equilibrium point by a disturbing force, the spring will actually deform and remain stretched out, instead of returning to the equilibrium position. This caveat to Hooke's law is particularly important for those who incorporate an oscillator into the design of an amusement park ride. If the displacement exceeds the allowable range, the restoring force can no longer return the oscillator to equilibrium. For a bungee ride, this issue can pose safety risks, because the riders would continue to fall toward the ground. It is therefore very important to know how much mass can be added to a bungee cord before it deforms or breaks. One way to determine the range of mass that can safely be added to a bungee is to use a smaller-scale model of the system to explore the relationship between mass and displacement. In addition, we can use a rubber band to approximate the behavior of a bungee cord because rubber bands are much smaller and much less expensive. This type of modeling allows engineers to determine parameters they will need to consider when doing tests using actual bungee cords.

Your Task

Use what you know about forces, oscillation, systems and system models, and stability and change to design and carry out an investigation to determine the range of mass that can be added to a bungee cord so that it still obeys Hooke's law.

The guiding question of this investigation is, *Under what conditions do rubber bands obey Hooke's law?*

Materials

You may use any of the following materials during your investigation:

Consumables
- Rubber bands
- Tape

Equipment
- Safety glasses or goggles (required)
- Hanging mass set
- Paper clips
- Ruler
- Support stand
- Electronic or triple beam balance
- Stopwatch

If you have access to the following equipment, you may also consider using a video camera and a computer or tablet with video analysis software.

Safety Precautions

Follow all normal lab safety rules. In addition, take the following safety precautions:

1. Wear sanitized safety glasses or goggles during lab setup, hands-on activity, and takedown.

2. Keep fingers and toes out of the way of moving objects.

3. Wash hands with soap and water after completing the lab.

LAB 15

Investigation Proposal Required? ☐ Yes   ☐ No

Getting Started

To answer the guiding question, you will need to design and carry out an investigation to determine the range of mass that can be added to a rubber band so that it still obeys Hooke's law. Figure L15.2 illustrates how you can use the available equipment to study the motion of a mass–rubber band system. Before you can design your investigation, however, you must determine what type of data you need to collect, how you will collect it, and how you will analyze it.

To determine *what type of data you need to collect*, think about the following questions:

- What are the boundaries and components of the mass–rubber band system?
- How do the components of the system interact with each other?
- When is this system stable, and under which conditions does it change?
- How could you keep track of changes in this system quantitatively?
- How will you measure the motion of the hanging mass?
- What are the independent variables and the dependent variables for each experiment?

To determine *how you will collect the data*, think about the following questions:

- What other factors will you need to control during each experiment?
- What scale or scales should you use when you take your measurements?
- How will you make sure that your data are of high quality (i.e., how will you reduce error)?
- How will you keep track of and organize the data you collect?

To determine *how you will analyze the data*, think about the following questions:

- What types of patterns might you look for as you analyze your data?
- What type of table or graph could you create to help make sense of your data?
- How will you model the system to indicate under what parameters the harmonic motion is stable?

FIGURE L15.2

One way to examine the motion of the mass–rubber band system using the available equipment

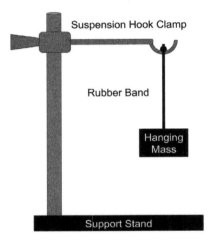

Suspension Hook Clamp

Rubber Band

Hanging Mass

Support Stand

Connections to the Nature of Scientific Knowledge and Scientific Inquiry

As you work through your investigation, you may want to consider

- how the culture of science, societal needs, and current events influence the work of scientists; and

- the role that imagination and creativity play in scientific research.

Initial Argument

Once your group has finished collecting and analyzing your data, your group will need to develop an initial argument. Your initial argument needs to include a claim, evidence to support your claim, and a justification of the evidence. The *claim* is your group's answer to the guiding question. The *evidence* is an analysis and interpretation of your data. Finally, the *justification* of the evidence is why your group thinks the evidence matters. The justification of the evidence is important because scientists can use different kinds of evidence to support their claims. Your group will create your initial argument on a whiteboard. Your whiteboard should include all the information shown in Figure L15.3.

FIGURE L15.3

Argument presentation on a whiteboard

The Guiding Question:	
Our Claim:	
Our Evidence:	Our Justification of the Evidence:

Argumentation Session

The argumentation session allows all of the groups to share their arguments. One or two members of each group will stay at the lab station to share that group's argument, while the other members of the group go to the other lab stations to listen to and critique the other arguments. This is similar to what scientists do when they propose, support, evaluate, and refine new ideas during a poster session at a conference. If you are presenting your group's argument, your goal is to share your ideas and answer questions. You should also keep a record of the critiques and suggestions made by your classmates so you can use this feedback to make your initial argument stronger. You can keep track of specific critiques and suggestions for improvement that your classmates mention in the space below.

Critiques about our initial argument and suggestions for improvement:

LAB 15

If you are critiquing your classmates' arguments, your goal is to look for mistakes in their arguments and offer suggestions for improvement so these mistakes can be fixed. You should look for ways to make your initial argument stronger by looking for things that the other groups did well. You can keep track of interesting ideas that you see and hear during the argumentation in the space below. You can also use this space to keep track of any questions that you will need to discuss with your team.

Interesting ideas from other groups or questions to take back to my group:

Once the argumentation session is complete, you will have a chance to meet with your group and revise your initial argument. Your group might need to gather more data or design a way to test one or more alternative claims as part of this process. Remember, your goal at this stage of the investigation is to develop the best argument possible.

Report

Once you have completed your research, you will need to prepare an *investigation report* that consists of three sections. Each section should provide an answer to the following questions:

1. What question were you trying to answer and why?

2. What did you do to answer your question and why?

3. What is your argument?

Your report should answer these questions in two pages or less. This report must be typed, and any diagrams, figures, or tables should be embedded into the document. Be sure to write in a persuasive style; you are trying to convince others that your claim is acceptable or valid!

Checkout Questions

Lab 15. Simple Harmonic Motion and Rubber Bands: Under What Conditions Do Rubber Bands Obey Hooke's Law?

1. What is the mathematical relationship between the force acting on the rubber band and the elongation of the rubber band?

   Is the function linear as the force increases?

2. In springs, the spring constant *k* is a function of both the material the spring is made from and the shape of the spring. What factors do you think might affect the constant of proportionality relating the force on a rubber band to the elongation of the rubber band?

Explain your answer, based on what you observed during your investigation about Hooke's law and rubber bands.

3. The imagination and creativity of a scientist play an important role in planning and carrying out investigations.

    a. I agree with this statement.
    b. I disagree with this statement.

Explain your answer, using an example from your investigation about Hooke's law and rubber bands.

4. The research done by a scientist is often influenced by what is important in society.

    a. I agree with this statement.
    b. I disagree with this statement.

Explain your answer, using an example from your investigation about Hooke's law and rubber bands.

5. Models are used to understand complex phenomena across the different scientific disciplines. Explain why models are so important, using an example from your investigation about Hooke's law and rubber bands.

6. Scientists often seek to identify the parameters under which a system is stable and what happens to the system when those parameters are exceeded. Explain why this is such an important research aim, using an example from your investigation about Hooke's law and rubber bands.

SECTION 6
Forces and Motion

Systems of Particles and Linear Momentum

Introduction Labs

LAB 16

Lab Handout

Lab 16. Linear Momentum and Collisions: When Two Objects Collide and Stick Together, How Do the Initial Velocity and Mass of One of the Moving Objects Affect the Velocity of the Two Objects After the Collision?

Introduction

The incidence of traumatic brain injury (TBI) is on the rise in the United States (CDC 2016). At least 1.7 million TBIs occur every year in this country, and these injuries are a contributing factor in about a third (30.5%) of all injury-related deaths (Faul et al. 2010). Adolescents (ages 15–19 years), older adults (ages 65 years and older), and males across all age groups are most likely to sustain a TBI (Faul et al. 2010). A TBI is caused by a bump, blow, or jolt to the head, although not all such events result in a TBI. TBIs can range from mild to severe, but most TBIs are mild and are commonly called concussions (CDC 2016). A single concussion, however, can cause temporary memory loss, confusion, and impaired vision or hearing. It can also cause depression, anxiety, and mood swings. Multiple concussions can make these symptoms permanent.

A concussion is caused when a person's brain hits the inside of the skull. This can occur when a moving person hits a stationary object (like running into a glass door), when a stationary person is hit by a moving object (like a person getting hit in the head by a baseball), or when two people collide with each other (like two people running into each other while playing a sport). As can be seen in Figure L16.1, when the skull is jolted too fast or is impacted by something, the brain shifts and hits against the skull. The "harder" the brain collides with the inside of the skull, the more severe the concussion.

Scientists have shown that the momentum of a collision is related to the severity of a concussion. Consider for example, what happens when a person is hit in the head with a ball. This often happens to softball players, baseball players, and soccer players. When a person is hit in the head by a moving ball, the severity of the concussion will be related to the momentum of that moving ball. Momentum is a function of both the mass of an object and its velocity. As an object accelerates, its

FIGURE L16.1

Movement of the brain during a concussive event

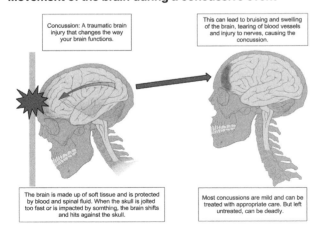

Concussion: A traumatic brain injury that changes the way your brain functions.

This can lead to bruising and swelling of the brain, tearing of blood vessels and injury to nerves, causing the concussion.

The brain is made up of soft tissue and is protected by blood and spinal fluid. When the skull is jolted too fast or is impacted by somthing, the brain shifts and hits against the skull.

Most concussions are mild and can be treated with appropriate care. But left untreated, can be deadly.

Linear Momentum and Collisions

When Two Objects Collide and Stick Together, How Do the Initial Velocity and Mass of One of the Moving Objects Affect the Velocity of the Two Objects After the Collision?

momentum increases. And, for two objects moving at the same velocity, the object with the greater mass will have a greater momentum.

Scientists have been studying collisions between two objects, such as cars, for some time. More and more scientists, however, are now studying the types of collisions that happen during different kinds of sports to better protect athletes from concussions. One type of collision that appears to be related to a high incidence of concussion is tackling in football. Tackling often results in two bodies staying together after the collision; Figure L16.2 shows one example of this. In this example, a defensive player (who is moving) collides with the quarterback (who is stationary). The defensive player holds on to the quarterback after the initial collision so they stay together as they fall to the ground. One goal of this line of research is to determine how the velocity of the moving object (or an athlete) before the collision affects the velocity of the two objects stuck together after the collision. In this investigation, you will have an opportunity to explore this relationship.

FIGURE L16.2 _____

A collision between two football players

Your Task

Use what you know about momentum, collisions, systems, tracking the movement of matter within systems, and the importance of considering issues related to scale, proportion, and quantity to design and carry out an investigation that will allow you to understand what happens when two objects collide and stick together.

The guiding question of this investigation is, *When two objects collide and stick together, how does the initial velocity and mass of one of the moving objects affect the velocity of the two objects after the collision?*

Materials

You may use any of the following materials during your investigation (some items may not be available):

- Safety glasses or goggles (required)
- 2 Dynamics carts (with Velcro or magnetic bumpers)
- Dynamics track
- Motion detector/sensor and interface
- Video camera
- Computer or tablet with data collection and analysis software and/or video analysis software

- Electronic or triple beam balance
- Cart picket fence
- Mass set
- Stopwatch
- Meterstick or ruler

Safety Precautions

Follow all normal lab safety rules. In addition, take the following safety precautions:

1. Wear sanitized safety glasses or goggles during lab setup, hands-on activity, and takedown.

2. Keep fingers and toes out of the way of moving objects.

3. Wash your hands with soap and water after completing the lab.

Investigation Proposal Required?   ☐ Yes   ☐ No

Getting Started

To answer the guiding question, you will need to design and carry out at least two different experiments. First, you will need to determine how changing the initial velocity of a moving object affects the velocity of the two objects after the collision. Next, you will need to determine how changing the mass of the moving object affects the velocity of the two objects after the collision. Figure L16.3 shows how you can use motion detectors/sensors to measure the velocity of a moving object (in this case, a dynamics cart) before and after a collision. The velocity of the moving object can also be measured by using a video camera and video analysis software. Before you can design your two experiments, however, you must determine what type of data you need to collect, how you will collect it, and how you will analyze it.

FIGURE L16.3 _____

One way to measure the velocity of a moving object before and after a collision

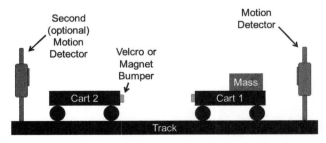

To determine *what type of data you need to collect,* think about the following questions:

- What are the boundaries and components of the system you are studying?
- How do components of the system under study interact?
- How will you track the movement of matter within this system?
- How could you keep track of changes in this system quantitatively?
- What factors affects the momentum of an object?
- How will you determine the velocity of each object?
- What will be the independent variable and the dependent variable for each experiment?

To determine *how you will collect the data,* think about the following questions:

- What other factors will you need to control or measure during each experiment?
- Which quantities are vectors, and which quantities are scalars?
- For any vector quantities, which directions are positive and which directions are negative?
- What scale or scales should you use to take your measurements?
- What equipment will you need to collect the measurements you need?
- How will you make sure that your data are of high quality (i.e., how will you reduce error)?
- How will you keep track of and organize the data you collect?

To determine *how you will analyze the data,* think about the following questions:

- What type of calculations will you need to do?
- What types of patterns might you look for as you analyze your data?
- Are there any proportional relationships you can identify?
- What types of comparisons will be useful to make?
- What type of table or graph could you create to help make sense of your data?

Connections to the Nature of Scientific Knowledge and Scientific Inquiry

As you work through your investigation, you may want to consider

- the difference between observations and inferences in science, and
- how the culture of science, societal needs, and current events influence the work of scientists.

LAB 16

Initial Argument

Once your group has finished collecting and analyzing your data, your group will need to develop an initial argument. Your initial argument needs to include a claim, evidence to support your claim, and a justification of the evidence. The *claim* is your group's answer to the guiding question. The *evidence* is an analysis and interpretation of your data. Finally, the *justification* of the evidence is why your group thinks the evidence matters. The justification of the evidence is important because scientists can use different kinds of evidence to support their claims. Your group will create your initial argument on a whiteboard. Your whiteboard should include all the information shown in Figure L16.4.

FIGURE L16.4

Argument presentation on a whiteboard

The Guiding Question:	
Our Claim:	
Our Evidence:	Our Justification of the Evidence:

Argumentation Session

The argumentation session allows all of the groups to share their arguments. One or two members of each group will stay at the lab station to share that group's argument, while the other members of the group go to the other lab stations to listen to and critique the other arguments. This is similar to what scientists do when they propose, support, evaluate, and refine new ideas during a poster session at a conference. If you are presenting your group's argument, your goal is to share your ideas and answer questions. You should also keep a record of the critiques and suggestions made by your classmates so you can use this feedback to make your initial argument stronger. You can keep track of specific critiques and suggestions for improvement that your classmates mention in the space below.

Critiques about our initial argument and suggestions for improvement:

If you are critiquing your classmates' arguments, your goal is to look for mistakes in their arguments and offer suggestions for improvement so these mistakes can be fixed. You should look for ways to make your initial argument stronger by looking for things that the other groups did well. You can keep track of interesting ideas that you see and hear during the argumentation in the space below. You can also use this space to keep track of any questions that you will need to discuss with your team.

Interesting ideas from other groups or questions to take back to my group:

Once the argumentation session is complete, you will have a chance to meet with your group and revise your initial argument. Your group might need to gather more data or design a way to test one or more alternative claims as part of this process. Remember, your goal at this stage of the investigation is to develop the best argument possible.

Report

Once you have completed your research, you will need to prepare an *investigation report* that consists of three sections. Each section should provide an answer to the following questions:

1. What question were you trying to answer and why?

2. What did you do to answer your question and why?

3. What is your argument?

LAB 16

Your report should answer these questions in two pages or less. This report must be typed, and any diagrams, figures, or tables should be embedded into the document. Be sure to write in a persuasive style; you are trying to convince others that your claim is acceptable or valid!

References

Centers for Disease Control and Prevention. 2016. TBI: Get the facts. *www.cdc.gov/traumaticbraininjury/get_the_facts.html.*

Faul, M., L. Xu, M. M. Wald, and V. G. Coronado. 2010. Traumatic brain injury in the United States: Emergency department visits, hospitalizations and deaths. Atlanta, GA: Centers for Disease Control and Prevention, National Center for Injury Prevention and Control.

Linear Momentum and Collisions

When Two Objects Collide and Stick Together, How Do the Initial Velocity and Mass of One of the Moving Objects Affect the Velocity of the Two Objects After the Collision?

Checkout Questions

Lab 16. Linear Momentum and Collisions: When Two Objects Collide and Stick Together, How Do the Initial Velocity and Mass of One of the Moving Objects Affect the Velocity of the Two Objects After the Collision?

The images below show the motion of two carts on a track before they collide with each other. Assume that both carts stick together after the collision. Use this information to answer questions 1 and 2.

1. How would the magnitude of the velocity of the carts after the collision in situation A compare with the magnitude of the velocity of the carts after the collision in situation B? For situation B, assume the magnitude of the velocity for cart 1 equals the magnitude of the velocity for cart 2.

    a. The velocity will be greater in A than in B.

    b. The velocity will be less in A than in B.

    c. The velocity will be equal in A and B.

(A)

(B)

(C)

How do you know?

2. How would the magnitude of the velocity of the carts after the collision in situation A compare with the magnitude of the velocity of the carts after the collision in situation C? For situation C, assume the magnitude of the velocity for cart 1 is greater than the magnitude of the velocity for cart 2.

    a. The velocity will be greater in A than in C.

    b. The velocity will be less in A than in C.

    c. The velocity will be equal in A and C.

How do you know?

3. The mass of the carts did not change while they were moving during your investigation. Are there instances where the mass of a moving object changes as it moves?

    a. Yes

    b. No

Explain your answer using an example.

4. How does decreasing the mass of a moving object as it moves affect the momentum of that object?

    a. It decreases the momentum of the object.

    b. It increases the momentum of the object.

    c. It has no effect on the momentum of the object.

How do you know?

5. In science, there is a difference between inferences and observations.

   a. I agree with this statement.
   b. I disagree with this statement.

   Explain your answer, using an example from your investigation about linear momentum and collisions.

6. Scientists share a set of values, norms, and commitments that shape what counts as knowing, how to represent or communicate information, and how to interact with other scientists.

   a. I agree with this statement.
   b. I disagree with this statement.

   Explain your answer, using an example from your investigation about linear momentum and collisions.

LAB 16

7. Scientists often need to need to define the system under study as part of the investigation. Explain why this is useful to do, using an example from your investigation about linear momentum and collisions.

8. Scientists often need to track how matter moves within a system. Explain why this is useful to do, using an example from your investigation about linear momentum and collisions.

9. Scientists often focus on proportional relationships. Explain what a proportional relationship is and why these relationships are useful, using an example from your investigation about linear momentum and collisions.

Lab Handout

Lab 17. Impulse and Momentum: How Does Changing the Magnitude and Duration of a Force Acting on an Object Affect the Momentum of That Object?

Introduction

Forces are responsible for all changes in motion and momentum. Regardless of how quickly a force is applied, it can still change the motion of an object. Consider what happens when someone hits a ball with a bat. The time that the bat and ball are in contact is very short, but the force from this collision is strong enough to significantly change the motion of a ball. The force that results from a bat hitting a ball can even be strong enough to change the shape of the ball (see Figure L17.1) or the bat (see Figure L17.2).

FIGURE L17.1 _____

A ball deforming from the force that results from the collision of the bat and the ball

FIGURE L17.2 _____

The bat breaking from the force that results from the collision of the bat and the ball

Momentum is defined as the mass of an object multiplied by its velocity. Momentum is a vector quantity because it has both a magnitude and a direction. As a result, the momentum of an object can be positive or negative, depending on the direction an object is moving. Force is also a vector quantity because forces have both a magnitude and a direction. When an unbalanced force acts on an object, the momentum of that object will change. In other words, an unbalanced force will change the momentum of an object.

LAB 17

The amount of time that an unbalanced force acts on an object is also important to consider when examining the change in momentum of an object. Sometimes the amount of time a force is applied to an object is very short, such as when a bat hits a ball, and other times it is applied over long periods, such as when the thrusters attached to a satellite are fired for several minutes to launch that satellite into orbit. The term *impulse* is used to describe the product of the magnitude and the duration of a force that acts on an object. In this investigation you will have an opportunity to examine how the nature of an impulse can change the momentum of a cart moving in one dimension. Your goal is to create a conceptual model that you can use to explain how the magnitude and duration of a force affects the change in momentum of a cart.

Your Task

Use what you know about momentum, impulse, the movement of matter within a system, and scale, proportional relationships, and quantity to develop a conceptual model that will enable you to explain how the momentum of an object will change in response to an impulse. To develop this conceptual model, you will need to design and carry out two different experiments to determine how (a) the magnitude of a force affects the momentum of an object and (b) the duration of a force affects the momentum of an object. Once you have developed your model, you will need to test it to determine if allows you to make accurate predictions about the change of momentum of an object over time in response to different types of impulse.

The guiding question of this investigation is, *How does changing the magnitude and the duration of a force acting on an object affect the momentum of that object?*

Materials

You may use any of the following materials during your investigation (some items may not be available):

- Safety glasses or goggles (required)
- Dynamics cart with fan attachment
- Dynamics track
- Motion detector/sensor and interface
- Video camera
- Computer or tablet with data collection and analysis software and/ or video analysis software
- Electronic or triple beam balance
- Stopwatch
- Meterstick or ruler

Safety Precautions

Follow all normal lab safety rules. In addition, take the following safety precautions:

1. Wear sanitized safety glasses or goggles during lab setup, hands-on activity, and takedown.

2. Keep fingers and toes out of the way of the moving objects.

3. Wash hands with soap and water after completing the lab.

Investigation Proposal Required?   ☐ Yes   ☐ No

Getting Started

The first step in developing your conceptual model is to design and carry out two experiments. In the first experiment, you will need to determine how changing the magnitude of a force will affect the momentum of a cart. In the second experiment, you will need to determine how changing the duration of the force affects the momentum of a cart. Figure L17.3 illustrates how you can use the available equipment to study the momentum of cart moving in one dimension. Before you can design your experiments, however, you must determine what type of data you need to collect, how you will collect it, and how you will analyze it.

FIGURE L17.3

One way to study the change in momentum of an object

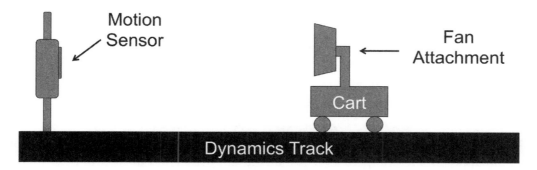

To determine *what type of data you need to collect,* think about the following questions:

- What are the boundaries and components of the system you are studying?
- How do components of the system under study interact?
- How will you track the movement of matter within this system?
- How could you keep track of changes in this system quantitatively?
- What factors affects the momentum of an object?
- How will you determine the velocity of each object?
- What will be the independent variable and the dependent variable for each experiment?

LAB 17

To determine *how you will collect the data*, think about the following questions:

- What other factors will you need to control or measure during each experiment?
- Which quantities are vectors, and which quantities are scalars?
- For any vector quantities, which directions are positive and which directions are negative?
- What scale or scales should you use when you take your measurements?
- What equipment will you need to collect the measurements you need?
- How will you make sure that your data are of high quality (i.e., how will you reduce error)?
- How will you keep track of and organize the data you collect?

To determine *how you will analyze the data*, think about the following questions:

- What type of calculations will you need to do?
- What types of patterns might you look for as you analyze your data?
- Are there any proportional relationships you can identify?
- What types of comparisons will be useful to make?
- What type of table or graph could you create to help make sense of your data?

Once you have determined how the magnitude of a force and the duration of a force affect the momentum of a cart in one dimension, your group will need to develop a conceptual model. Your model must include the various forces acting on the cart and allow you to make accurate predictions about how the momentum of cart changes over time in response to different forces.

The last step in this investigation will be to test your model. To accomplish this goal, you can apply different impulses (ones that you did not test) to the cart to determine if your model enables you to make accurate predictions about how the momentum of the cart changes over time. If you are able to use your model to make accurate predictions, then you will be able to generate the evidence you need to convince others that your model is a valid and acceptable. The fan attached to the cart you will use in this investigation may have a limited number of different speeds, so it will be important to reserve at least one speed setting for this step of your investigation.

Connections to the Nature of Scientific Knowledge and Scientific Inquiry

As you work through your investigation, you may want to consider

- the difference between laws and theories in science, and
- the difference between data and evidence in science.

Initial Argument

Once your group has finished collecting and analyzing your data, your group will need to develop an initial argument. Your initial argument needs to include a claim, evidence to support your claim, and a justification of the evidence. The *claim* is your group's answer to the guiding question. The *evidence* is an analysis and interpretation of your data. Finally, the *justification* of the evidence is why your group thinks the evidence matters. The justification of the evidence is important because scientists can use different kinds of evidence to support their claims. Your group will create your initial argument on a whiteboard. Your whiteboard should include all the information shown in Figure L17.4.

FIGURE L17.4 _____

Argument presentation on a whiteboard

The Guiding Question:	
Our Claim:	
Our Evidence:	Our Justification of the Evidence:

Argumentation Session

The argumentation session allows all of the groups to share their arguments. One or two members of each group will stay at the lab station to share that group's argument, while the other members of the group go to the other lab stations to listen to and critique the other arguments. This is similar to what scientists do when they propose, support, evaluate, and refine new ideas during a poster session at a conference. If you are presenting your group's argument, your goal is to share your ideas and answer questions. You should also keep a record of the critiques and suggestions made by your classmates so you can use this feedback to make your initial argument stronger. You can keep track of specific critiques and suggestions for improvement that your classmates mention in the space below.

Critiques about our initial argument and suggestions for improvement:

LAB 17

If you are critiquing your classmates' arguments, your goal is to look for mistakes in their arguments and offer suggestions for improvement so these mistakes can be fixed. You should look for ways to make your initial argument stronger by looking for things that the other groups did well. You can keep track of interesting ideas that you see and hear during the argumentation in the space below. You can also use this space to keep track of any questions that you will need to discuss with your team.

Interesting ideas from other groups or questions to take back to my group:

Once the argumentation session is complete, you will have a chance to meet with your group and revise your initial argument. Your group might need to gather more data or design a way to test one or more alternative claims as part of this process. Remember, your goal at this stage of the investigation is to develop the best argument possible.

Report

Once you have completed your research, you will need to prepare an *investigation report* that consists of three sections. Each section should provide an answer to the following questions:

1. What question were you trying to answer and why?

2. What did you do to answer your question and why?

3. What is your argument?

Your report should answer these questions in two pages or less. This report must be typed, and any diagrams, figures, or tables should be embedded into the document. Be sure to write in a persuasive style; you are trying to convince others that your claim is acceptable or valid!

Checkout Questions

Lab 17. Impulse and Momentum: How Does Changing the Magnitude and Duration of a Force Acting on an Object Affect the Momentum of That Object?

1. How can the shape of a force versus time graph be used to determine an object's momentum?

Use the following information to answer questions 2–4. Consider a cart starting from rest with a fan attachment that applies a constant force. Assume that there is no friction acting on the cart as it moves.

2. What would the momentum versus time graph look like if the fan force doubled halfway through the trial?

A

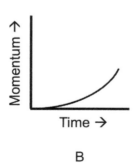
B

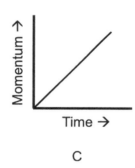
C

a. Graph A

b. Graph B

c. Graph C

How do you know?

LAB 17

3. What could cause the slope to become zero?

    a. Doubling the fan force a second time
    b. Turning the fan off
    c. Leaving the fan as is

    How do you know?

4. Draw a momentum versus time graph showing the change in momentum if the fan stayed on for twice as long.

    Why did you draw your momentum versus time graph like that?

5. There is a difference between a scientific law and a scientific theory.

    a. I agree with this statement.
    b. I disagree with this statement.

    Explain your answer, using an example from your investigation about impulse and momentum.

6. There is a difference between data and evidence in science.

   a. I agree with this statement.
   b. I disagree with this statement.

   Explain your answer, using an example from your investigation about impulse and momentum.

7. In physics, it is important to classify something as either a vector quantity or a scalar quantity. Explain what a vector is and why it is important to identify vector quantities in physics, using an example from your investigation about impulse and momentum.

8. Scientists often need to track how energy or matter moves into, out of, or within a system during an investigation. Explain why tracking energy and matter is such an important part of science, using an example from your investigation about impulse and momentum.

Application Labs

LAB 18

Lab Handout

Lab 18. Elastic and Inelastic Collisions: Which Properties of a System Are Conserved During a Collision?

Introduction

Physics is the scientific study of time, space, and matter. Some branches of physics, such as cosmology, investigate questions regarding the entire universe (e.g., how old is it, how did it begin). Most branches of physics, however, investigate questions related to smaller scales and systems. When studying a system, a physicist will identify the system and then ask questions about (1) the matter contained in that system, (2) the interactions between matter contained in the system, and (3) how the matter moves in the system. When doing this, the physicist ignores any influences from outside the system during the investigation, while recognizing that those influences are still there. For example, when studying acceleration of an object in free fall due to gravity, a physicist might ignore the influence of air resistance when its effects are less than the uncertainty in the measurements. Systems come in all sizes. Astronomers study systems as large as galaxies. Chemists study systems that can be as small as a few atoms.

Scientists have identified several laws of conservation that are the same across all systems. A conservation law states that some property of an isolated system is the same before and after some specific interaction takes place within that system. The properties of each object in the system, however, do not need to be the same before and after the interaction in order for some property of the system to stay the same. For example, the law of conservation of energy indicates that the total amount of energy in a system stays the same before and after any interaction that takes place between one or more components of that system. To illustrate what the conservation of energy means, consider what happens when a person places a hot metal spoon into a cold cup of water. When the hot metal spoon is placed into the cup, some heat energy will transfer from the spoon into the water, but the total amount of energy of the system remains constant because the energy that transferred from the spoon into the water is still a part of the system. The transfer of energy caused the temperature of the water to increase and the temperature of the metal to decrease but the total amount of energy in the system did not change, so energy is conserved within the system.

When studying the interactions between two or more objects in a system, physicists often try to identify which properties are conserved during the interaction. In the example of putting a hot metal spoon into cold water, energy is conserved but temperature is not. Another type of interaction that physicists often study is a collision. Collisions are a common experience, from billiard balls colliding on a pool table to an asteroid hitting a planet, or, as shown in Figure L18.1, a collision between two cars.

There are a number of properties that could be conserved during a collision. Some examples include acceleration, velocity, force, energy, and momentum. There are likely other properties that might also be conserved during a collision. In this investigation, you will have an opportunity to determine which properties of a system are conserved during a two-car collision.

FIGURE L18.1 _____

A collision between two vehicles

Your Task

Use what you know about momentum, velocity, acceleration, the conservation of energy and matter, and systems and system models to design and carry out an investigation that will allow you to understand what happens to the different properties of a two-car system before and after a collision.

The guiding question of this investigation is, *Which properties of a system are conserved during a collision?*

Materials

You may use any of the following materials during your investigation:

- Safety glasses or goggles (required)
- Dynamics carts
- Dynamics track
- Bumper kit for the carts
- Video camera
- Computer or tablet with video analysis software
- Electronic or triple beam balance
- Mass set
- Stopwatches
- Ruler

Safety Precautions

Follow all normal lab safety rules. In addition, take the following safety precautions:

1. Wear sanitized safety glasses or goggles during lab setup, hands-on activity, and takedown.

2. Keep fingers and toes out of the way of moving objects.

3. Wash hands with soap and water after completing the lab.

Investigation Proposal Required?   ☐ Yes   ☐ No

LAB 18

Getting Started

To answer the guiding question you will need to design and carry out two experiments. Figure L18.2 shows how you can set up two carts on a track to examine changes in the velocity, acceleration, and position of each cart before and after a collision. You can also change the nature of the collision by changing the bumpers on the carts. You can use a hoop bumper to make the carts bounce apart after they collide or a Velcro or magnet bumper to make them stick together. To measure changes in velocity, acceleration, and the position of the two carts at the same time, you will need to use a video camera and video analysis software. Before you can design your two experiments, however, you must first determine what type of data you need to collect, how you will collect it, and how you will analyze it.

FIGURE L18.2

One way to measure the velocity, acceleration, or position of a moving object before and after a collision

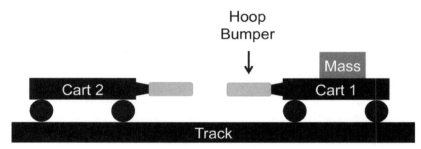

To determine *what type of data you need to collect,* think about the following questions:

- What are the boundaries and components of the system under study?
- How do the components of the system interact?
- What properties of the system might be conserved?
- What properties of the system are directly measurable?
- What properties of the system will you need to calculate from other measurements?
- What types of collisions will you need to model?
- How can you track how matter and energy flows into, out of, or within this system?
- What will be the independent variable and the dependent variable for each experiment?

To determine *how you will collect the data,* think about the following questions:

- What other factors will you need to control during each experiment?

- Which quantities are vectors, and which quantities are scalars?

- What scale or scales should you use when you take your measurements?

- What equipment will you need to collect the data you need?

- How will you make sure that your data are of high quality (i.e., how will you reduce error)?

- How will you keep track of and organize the data you collect?

To determine *how you will analyze the data,* think about the following questions:

- How will you determine if a property has been conserved during a collision?

- What type of calculations will you need to make?

- What types of comparison will be useful for you to make?

- How could you use mathematics to describe a relationship between variables?

- What type of table or graph could you create to help make sense of your data?

Connections to the Nature of Scientific Knowledge and Scientific Inquiry

As you work through your investigation, you may want to consider

- the difference between laws and theories in science, and

- the difference between data and evidence in science.

Initial Argument

Once your group has finished collecting and analyzing your data, your group will need to develop an initial argument. Your initial argument needs to include a claim, evidence to support your claim, and a justification of the evidence. The *claim* is your group's answer to the guiding question. The *evidence* is an analysis and interpretation of your data. Finally, the *justification* of the evidence is why your group thinks the evidence matters. The justification of the evidence is important because scientists can use different kinds of evidence to support their claims. Your group will create your initial argument on a whiteboard. Your whiteboard should include all the information shown in Figure L18.3.

FIGURE L18.3 _____
Argument presentation on a whiteboard

The Guiding Question:	
Our Claim:	
Our Evidence:	Our Justification of the Evidence:

LAB 18

Argumentation Session

The argumentation session allows all of the groups to share their arguments. One or two members of each group will stay at the lab station to share that group's argument, while the other members of the group go to the other lab stations to listen to and critique the other arguments. This is similar to what scientists do when they propose, support, evaluate, and refine new ideas during a poster session at a conference. If you are presenting your group's argument, your goal is to share your ideas and answer questions. You should also keep a record of the critiques and suggestions made by your classmates so you can use this feedback to make your initial argument stronger. You can keep track of specific critiques and suggestions for improvement that your classmates mention in the space below.

Critiques about our initial argument and suggestions for improvement:

If you are critiquing your classmates' arguments, your goal is to look for mistakes in their arguments and offer suggestions for improvement so these mistakes can be fixed. You should look for ways you to make your initial argument stronger by looking for things that the other groups did well. You can keep track of interesting ideas that you see and hear during the argumentation in the space below. You can also use this space to keep track of any questions that you will need to discuss with your team.

Interesting ideas from other groups or questions to take back to my group:

National Science Teachers Association

Once the argumentation session is complete, you will have a chance to meet with your group and revise your initial argument. Your group might need to gather more data or design a way to test one or more alternative claims as part of this process. Remember, your goal at this stage of the investigation is to develop the best argument possible.

Report

Once you have completed your research, you will need to prepare an *investigation report* that consists of three sections. Each section should provide an answer to the following questions:

1. What question were you trying to answer and why?

2. What did you do to answer your question and why?

3. What is your argument?

Your report should answer these questions in two pages or less. This report must be typed, and any diagrams, figures, or tables should be embedded into the document. Be sure to write in a persuasive style; you are trying to convince others that your claim is acceptable or valid!

LAB 18

Lab 18. Elastic and Inelastic Collisions: Which Properties Are Conserved During a Collision?

1. In your investigation, the kinetic energy of the system before the collision was greater than the kinetic energy of the system after the collision. Did your investigation violate the law of conservation of energy?

   a. Yes, it violated the law of conservation of energy.
   b. No, it did not violate the law of conservation of energy.

   Explain why or why not.

   Use the following information to answer questions 2 and 3. A truck has a mass of 5,000 kg and is moving to the right at 10 m/s. A small car has a mass of 2,000 kg and is moving to the left at 20 m/s. The two vehicles collide head on.

2. Is this collision elastic or inelastic?

   a. Elastic
   b. Inelastic

   How do you know?

3. What is the total momentum of the two-vehicle system before the collision?

   a. 10,000 kg·m/s
   b. 40,000 kg·m/s
   c. 50,000 kg·m/s
   d. 90,000 kg·m/s

   How do you know?

4. The terms *data* and *evidence* mean the same thing in science.

   a. I agree with this statement.
   b. I disagree with this statement.

   Explain your answer, using an example from your investigation about collisions.

LAB 18

5. It is important to track how energy and matter move into, out of, and within a system and to determine if any of properties are conserved within the system.

   a. I agree with this statement.
   b. I disagree with this statement.

   Explain your answer, using an example from your investigation about collisions.

6. In science, there is a difference between a law and a theory. What is the difference between a law and a theory? Explain why this distinction is important, using an example from your investigation about collisions.

7. Scientists often need to identify the system under study before they start collecting data. Explain why defining the system under study is so important in science, using an example from your investigation about collisions.

LAB 19

Lab 19. Impulse and Materials: Which Material Is Most Likely to Provide the Best Protection for a Phone That Has Been Dropped?

Introduction

People throughout the United States use personal electronic devices. On any given day, a person might use a cell phone, smartwatch, laptop computer, or tablet in a number of different settings. Unfortunately, people often damage these devices by accidently dropping them or knocking them off a table. Most consumers, however, do not want to replace any of these devices on a regular basis because they tend to be expensive. This is one reason why so many people use a case to protect their different personal electronic devices. Many different companies sell cases that are designed to prevent phones, tablets, or laptops from breaking when they are dropped or knocked off a table.

When a person drops a cell phone, the force of gravity causes the phone to accelerate toward the ground at −9.8 m/s². Any object with a velocity has momentum, and as an object accelerates during free fall, its momentum increases. Upon reaching the ground, the ground exerts an upward, normal force that causes the momentum of the object to change. Figure L19.1 is an example of a momentum versus time graph for a cell phone that falls toward and then strikes the ground, bouncing upward once, and then coming to rest. Point A in the graph corresponds to the cell phone falling out of a person's hand. Between points B and C, the cell phone is hitting the ground, with the ground exerting a force on the cell phone, causing it to bounce back up. In other words, the segment of the graph between points B and C corresponds to the collision between the phone and the ground. At point D, the cell phone has reached the highest point of its bounce, and then at point E, the cell phone hits the ground and finally comes to rest at point F. The segment of the graph between E and F corresponds to the second collision between the phone and the ground.

Momentum versus time graph for a falling cell phone

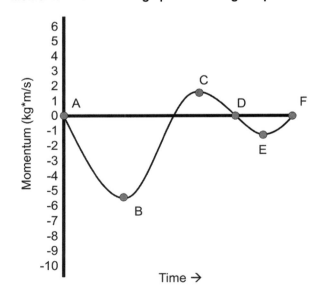

Cell phones tend to break easily when dropped because they are made out of lightweight materials, such as glass, aluminum, and plastic, which have a relatively low breaking point. The breaking point of a material is the maximum amount of force that it can absorb before it deforms. Companies make cell phones out of these materials because it allows the companies to make phones that are thin and light so they are easy to carry around. Unfortunately, a phone that is thin and light is also fragile.

Protective cases help alleviate this design trade-off. Protective phone cases tend to be made from a material that has two important characteristics. First, the material must be able to minimize the force that acts on the phone during a collision. This is important because the change of momentum associated with a phone in free fall suddenly hitting the ground is a vector that acts in the direction of the net force exerted on it. Second, the material must be able to maximize the time of the collision. This characteristic is important to consider because any change in momentum happens over a specific time interval. Your goal in this investigation is to use this information to identify a suitable material for a new protective cell phone case.

Your Task

Use what you know about momentum, impulses, the relationship between structure and function, and scale, proportional relationships, and vector quantities to design and carry out an investigation to compare how different materials change the momentum of an object during a collision.

The guiding question for this investigation is, **Which material is most likely to provide the best protection for a phone that has been dropped?**

Materials

You may use any of the following materials during your investigation:

- Safety glasses or goggles (required)
- Dynamics cart
- Dynamics track
- Force sensor
- Sensor interface
- Computer, tablet, or graphing calculator with data collection and analysis software
- Foam block
- Plastic block
- Rubber block
- Wooden block
- Metal block

Safety Precautions

Follow all normal lab safety rules. In addition, take the following safety precautions:

1. Wear sanitized safety glasses or goggles during lab setup, hands-on activity, and takedown.

2. Keep fingers and toes out of the way of moving objects.

LAB 19

3. Wash hands with soap and water after completing the lab.

Investigation Proposal Required? ☐ Yes   ☐ No

Getting Started

To answer the guiding question, you will need to investigate how different materials change the momentum of an object during a collision. Figure L19.2 shows how you can attach a force sensor to a cart and then roll it down a frictionless track on an incline. The end of the force sensor should collide with the material you are testing at the end of the track. Starting the cart at the same point on the track for each test will allow you to control for the momentum of the cart prior to the collision with the block. Before you can begin to design your investigation using this equipment, however, you must determine what type of data you need to collect, how you will collect it, and how you will analyze it.

FIGURE L19.2

One way to examine the change in momentum of the cart after a collision with a specific material

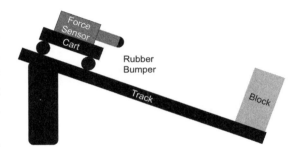

To determine *what type of data you need to collect*, think about the following questions:

- How might changes in the structure of a cell phone case affect the function of the case?
- What are the components of this system and how do they interact?
- How can you describe the components of the system quantitatively?
- How do you quantify a change in momentum?
- What measurements do you need to make?
- What will be the independent variable and the dependent variable for your experiment?

To determine *how you will collect the data*, think about the following questions:

- What other variables do you need to measure or control?
- What scale or scales should you use when you take your measurements?
- Which quantities are vectors, and which quantities are scalars?
- For any vector quantities, which directions are positive and which directions are negative?

- How will you make sure that your data are of high quality (i.e., how will you reduce error)?
- How will you keep track of and organize the data you collect?

To determine *how you will analyze the data,* think about the following questions:

- What type of calculations will you need to make?
- What types of comparisons will be useful?
- How could you use mathematics to document a difference between materials?
- What type of table or graph could you create to help make sense of your data?

Connections to the Nature of Scientific Knowledge and Scientific Inquiry

As you work through your investigation, you may want to consider

- how the culture of science, societal needs, and current events influence the work of scientists; and
- the nature and role of experiments in science.

Initial Argument

Once your group has finished collecting and analyzing your data, your group will need to develop an initial argument. Your initial argument needs to include a claim, evidence to support your claim, and a justification of the evidence. The *claim* is your group's answer to the guiding question. The *evidence* is an analysis and interpretation of your data. Finally, the *justification* of the evidence is why your group thinks the evidence matters. The justification of the evidence is important because scientists can use different kinds of evidence to support their claims. Your group will create your initial argument on a whiteboard. Your whiteboard should include all the information shown in Figure L19.3.

FIGURE L19.3 _____

Argument presentation on a whiteboard

The Guiding Question:	
Our Claim:	
Our Evidence:	Our Justification of the Evidence:

Argumentation Session

The argumentation session allows all of the groups to share their arguments. One or two members of each group will stay at the lab station to share that group's argument, while the other members of the group go to the other lab stations to listen to and critique the other arguments. This is similar to what scientists do when they propose, support, evaluate, and refine new ideas during a poster session at a conference. If you are presenting your group's argument, your goal is to share your ideas and answer questions. You should also keep a record of the critiques and suggestions made by your classmates so you can use this feedback to make your initial argument stronger. You can keep track of specific critiques and suggestions for improvement that your classmates mention in the space below.

Critiques about our initial argument and suggestions for improvement:

If you are critiquing your classmates' arguments, your goal is to look for mistakes in their arguments and offer suggestions for improvement so these mistakes can be fixed. You should look for ways to make your initial argument stronger by looking for things that the other groups did well. You can keep track of interesting ideas that you see and hear during the argumentation in the space below. You can also use this space to keep track of any questions that you will need to discuss with your team.

Interesting ideas from other groups or questions to take back to my group:

Once the argumentation session is complete, you will have a chance to meet with your group and revise your initial argument. Your group might need to gather more data or design a way to test one or more alternative claims as part of this process. Remember, your goal at this stage of the investigation is to develop the best argument possible.

Report

Once you have completed your research, you will need to prepare an *investigation report* that consists of three sections. Each section should provide an answer to the following questions:

1. What question were you trying to answer and why?

2. What did you do to answer your question and why?

3. What is your argument?

Your report should answer these questions in two pages or less. This report must be typed, and any diagrams, figures, or tables should be embedded into the document. Be sure to write in a persuasive style; you are trying to convince others that your claim is acceptable or valid!

LAB 19

Lab 19. Impulse and Materials: Which Material Is Most Likely to Provide the Best Protection for a Phone That Has Been Dropped?

1. How might your lab results regarding force, impulse, and momentum inform engineers when they design cars? How can this information help keep people safe during a collision between two cars?

2. Scientists and engineers who study ways to transport people to other planets must account for a number of challenges in the design of spaceships. One problem is how to get a large spaceship to move with a fast enough velocity. In response, some have suggested using a solar sail, where a ship uses a specially designed sail to catch microscopic particles continuously emitted by the Sun. How do the results of your lab relate to solar sails? Why might scientists think solar sails are a good solution?

3. How an object is structured is related to its function.

    a. I agree with this statement.
    b. I disagree with this statement.

    Explain your answer, using an example from your investigation about impulse and momentum.

4. The research done by a scientist is often influenced by current events or what is important in society.

    a. I agree with this statement.
    b. I disagree with this statement.

    Explain your answer, using an example from your investigation about impulse and momentum.

5. Experiments are one type of research design used by scientists. Explain what an experiment is and what types of questions are best answered using an experiment, using examples from your investigation about impulse and momentum as well as from previous investigations in this class or your previous science classes.

6. Scientists often need to think about scales, proportional relationships, and vector quantities during an investigation. Explain why it is important for scientists to think about these things, using examples from your investigation about impulse and momentum as well as from previous investigations in this class or your previous science classes.

SECTION 7
Energy, Work, and Power

Introduction Labs

LAB 20

Lab 20. Kinetic and Potential Energy: How Can We Use the Work-Energy Theorem to Explain and Predict Behavior of a System That Consists of a Ball, a Ramp, and a Cup?

Introduction

You have learned how to use Newton's laws of motion to explain how objects move and to predict the motion of an object over time. Newton's laws of motion therefore function as a useful model that people can use to understand how the world works and to predict how different objects move after they interact with each other. Physicists, however, can use other models to explain and predict the motion of objects and to help understand natural phenomena. One such model views the motion of objects through a lens of work and energy. This model is called the *work-energy theorem*. To understand how work and energy can be used to explain and predict motion, it is important to understand the basic assumptions underlying the work-energy theorem.

The first basic assumption of the work-energy theorem is that an object can store energy as the result of its position. This stored energy is called *potential energy*. The second basic assumption of the work-energy theorem is that all objects in motion have energy because they are moving. The energy of motion is called *kinetic energy*. There are many forms of kinetic energy, including vibrational (the energy due to vibrational motion), rotational (the energy due to rotational motion), and translational (the energy due to motion from one location to another). The third assumption of the work-energy theorem is that doing work on an object results in a change in the *mechanical energy* of that object. Mechanical energy is the sum of the potential and kinetic energy of an object. A change in mechanical energy of an object can result from adding energy to an object, taking away energy from an object, or changing the type of energy an object has from one form to another. Work is done on an object when a force acts on an object over a displacement. Therefore, *work* is mathematically defined as the product of the force acting on an object times the displacement. The fourth, and final, basic assumption of the work-energy theorem is that energy cannot be created or destroyed—it just changes form as objects move or as it transfers from one object to another one.

In this investigation you will have an opportunity to explore the relationship between potential energy, kinetic energy, mechanical energy, work, and displacement in terms of the motion of objects. To explore the relationship between these various components of the work-energy theorem, you will study a simple system that consists of a ball, a ramp, and a cup. You goal is to develop a set of rules that you can use to explain and predict the motion of the ball at the bottom of the ramp and the distance the cup moves after the ball rolls down the ramp and enters the cup. To be valid or acceptable, your set of rules must

Kinetic and Potential Energy
How Can We Use the Work-Energy Theorem to Explain and Predict Behavior of a System
That Consists of a Ball, a Ramp, and a Cup?

be consistent with the basic assumptions underlying the work-energy theorem outlined in this section.

Your Task

Use what you know about the work-energy theorem and the importance of tracking matter and energy in a system to develop a conceptual model (i.e., a set of rules) that will allow you to explain and predict the behavior of a system that consists of a ball, a ramp, and a cup. To develop your model, you will need to design and carry out several experiments to determine how changes in several different components of the ball-ramp-cup system affect (1) the motion of the ball at the bottom of the ramp after it rolls down it and (2) the distance a cup moves when a ball is rolled down an incline and comes into contact with the cup. Once you have developed your rules, you will need to test them to determine if you can use them to make accurate predictions about the behavior of the ball-ramp-cup system.

The guiding question of this investigation is, *How can we use the work-energy theorem to explain and predict behavior of a system that consists of a ball, a ramp, and a cup?*

Materials

You may use any of the following materials during your investigation:

- Safety glasses or goggles (required)
- Support stand
- Extension clamp
- Set of 25 mm balls (includes brass, aluminum, steel, cork, wood, and copper)
- PVC pipe
- Electronic or triple beam balance
- Plastic cup
- 2 Metersticks
- Protractor
- Stopwatch

If you have access to the following equipment, you may also consider using a video camera and a computer or tablet with video analysis software.

Safety Precautions

Follow all normal lab safety rules. In addition, take the following safety precautions:

1. Wear sanitized safety glasses or goggles during lab setup, hands-on activity, and takedown.

2. Keep fingers and toes out of the way of moving objects.

3. Wash hands with soap and water after completing the lab.

LAB 20

Investigation Proposal Required? ☐ Yes ☐ No

Getting Started

Your first step in this investigation is to design and carry several experiments to determine how changes in several different components of the ball-ramp-cup system affect (1) the motion of the ball at the bottom of the ramp after it rolls down it and (2) the distance a cup moves when a ball is rolled down an incline and comes into contact with the cup. Figure L20.1 shows how you can set up the ball-ramp-cup system. The components of the system that you can change include the mass of the ball (*m*), the height of the ramp (**h**), the length of the ramp (**l**), and the angle of inclination (θ). Before you design or carry out your experiments, however, you must first determine what type of data you need to collect, how you will collect it, and how you will analyze it.

FIGURE L20.1 _____

The ball-ramp-cup system

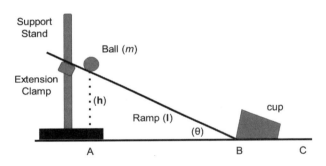

To determine *what type of data you need to collect*, think about the following questions:

- What are the boundaries and components of the system under study?
- How will you quantify the amount of potential, kinetic, and mechanical energy in the system?
- How can you track the potential, kinetic, and mechanical energy changes within this system?
- Which factors might control rates of change in the system under study?
- What will be the independent variable and the dependent variable for each experiment?

To determine *how you will collect your data*, think about the following questions:

- What other variables will you need to control in each experiment?
- What will be the reference point for measurement?

Kinetic and Potential Energy
*How Can We Use the Work-Energy Theorem to Explain and Predict Behavior of a System
That Consists of a Ball, a Ramp, and a Cup?*

- What measurement scale or scales should you use to collect data?
- What equipment will you need to take your various measurements?
- How will you make sure that your data are of high quality (i.e., how will you reduce error)?
- How will you keep track of and organize the data you collect?

To determine *how you will analyze your data*, think about the following questions:

- What type of calculations will you need to make?
- How could you use mathematics to describe a relationship between variables?
- What types of comparisons will be useful?
- What type of table or graph could you create to help make sense of your data?

Once you have determined how changes in the different components of the ball-ramp-cup system affect (1) the motion of the ball at the bottom of the ramp after it rolls down it and (2) the distance a cup moves when a ball is rolled down an incline and comes into contact with the cup, your group will need to develop a set of rules that you can use to explain and predict the behavior of this system. Your set of rules must be consistent with the four basic assumptions underlying the work-energy theorem outlined in the "Introduction."

The last step in this investigation will be to test your model. To accomplish this goal, you can set components of the system to values that you did not test (e.g., if you tested the height of the ramp at 0.5 m and 0.75 m, then set the height to 0.6 m) to determine if you can use your rulesto make accurate predictions, then you will be able to generate the evidence you need to convince others that your rules are a valid way to explain and predict behavior of the ball and the cup in terms of the work-energy theorem of motion.

Connections to the Nature of Scientific Knowledge and Scientific Inquiry

As you work through your investigation, you may want to consider

- how scientific knowledge changes over time, and
- the difference between laws and theories in science.

Initial Argument

Once your group has finished collecting and analyzing your data, your group will need to develop an initial argument. Your initial argument must include a claim, evidence to support your claim, and a *justification* of the evidence. The *claim* is your group's answer to the guiding question. The *evidence* is an analysis and interpretation of your data. Finally, the *justification* of the evidence is why your group thinks the evidence matters. The justification of the evidence is important because scientists can use different kinds of evidence to support their claims. You group will create your initial argument on a whiteboard. Your

LAB 20

whiteboard should include all the information shown in Figure L20.2.

Argumentation Session

The argumentation session allows all of the groups to share their arguments. One or two members of each group will stay at the lab station to share that group's argument, while the other members of the group go to the other lab stations to listen to and critique the other arguments. This is similar to what scientists do when they propose, support, evaluate, and refine new ideas during a poster session at a conference. If you are presenting your group's argument, your goal is to share your ideas and answer questions. You should also keep a record of the critiques and suggestions made by your classmates so you can use this feedback to make your initial argument stronger. You can keep track of specific critiques and suggestions for improvement that your classmates mention in the space below.

FIGURE L20.2 _____

Argument presentation on a whiteboard

The Guiding Question:	
Our Claim:	
Our Evidence:	Our Justification of the Evidence:

Critiques about our initial argument and suggestions for improvement:

If you are critiquing your classmates' arguments, your goal is to look for mistakes in their arguments and offer suggestions for improvement so these mistakes can be fixed. You should look for ways to make your initial argument stronger by looking for things that the other groups did well. You can keep track of interesting ideas that you see and hear during the argumentation in the space below. You can also use this space to keep track of any questions that you will need to discuss with your team.

Interesting ideas from other groups or questions to take back to my group:

Once the argumentation session is complete, you will have a chance to meet with your group and revise your initial argument. Your group might need to gather more data or design a way to test one or more alternative claims as part of this process. Remember, your goal at this stage of the investigation is to develop the best argument possible.

Report

Once you have completed your research, you will need to prepare an *investigation report* that consists of three sections. Each section should provide an answer to the following questions:

1. What question were you trying to answer and why?

2. What did you do to answer your question and why?

3. What is your argument?

Your report should answer these questions in two pages or less. This report must be typed, and any diagrams, figures, or tables should be embedded into the document. Be sure to write in a persuasive style; you are trying to convince others that your claim is acceptable or valid!

LAB 20

Lab 20. Kinetic and Potential Energy: How Can We Use the Work-Energy Theorem to Explain and Predict Behavior of a System That Consists of a Ball, a Ramp, and a Cup?

Use the figure below to answer questions 1 and 2. For the acceleration due to gravity, use the positive value for **g** (9.8 m/sec²).

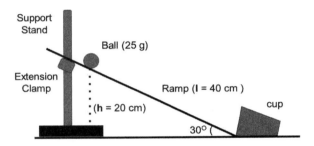

1. What is the potential energy of the ball at the moment it is released on the incline?

   How do you know?

2. What is the kinetic energy of the ball at the moment it strikes the cup?

   How do you know?

Kinetic and Potential Energy
*How Can We Use the Work-Energy Theorem to Explain and Predict Behavior of a System
That Consists of a Ball, a Ramp, and a Cup?*

3. Galileo hypothesized that free objects accelerate uniformly, or stated another way, that a falling object's velocity increases an equal amount in each equal time interval. Explain how the results of this ball and cup experiment could be used in support of this claim.

4. There is a difference between a law and a theory in science.

    a. I agree with this statement.
    b. I disagree with this statement.

    Explain your answer, using an example from your investigation about kinetic and potential energy.

LAB 20

5. Scientific knowledge, once proven to be true, does not change.

    a. I agree with this statement.

    b. I disagree with this statement.

Explain your answer, using an example from your investigation about kinetic and potential energy.

6. Scientists often need to identify a system and then create a model of it as part of an investigation. Explain why it useful to create models of systems, using an example from your investigation about kinetic and potential energy and an example from previous investigations in this class or your previous science classes.

Kinetic and Potential Energy
*How Can We Use the Work-Energy Theorem to Explain and Predict Behavior of a System
That Consists of a Ball, a Ramp, and a Cup?*

7. One of the important aims in science is to track how energy and matter move within a system and to determine if the energy and matter within the system are conserved. Explain why it is useful to track how energy and matter move within a system, using an example from your investigation about kinetic and potential energy and an example from previous investigations in this class or your previous science classes.

Lab Handout

Lab 21. Conservation of Energy and Pendulums: How Does Placing a Nail in the Path of a Pendulum Affect the Height of a Pendulum Swing?

Introduction

Two of the most influential thinkers in history were Aristotle in the 4th century BC and Galileo in the 16th–17th centuries. Aristotle took a philosophical approach to understanding the natural world, whereas Galileo preferred a more empirical one. Thus, they developed some very different explanations about the motion of objects. For example, Aristotle claimed that heavier bodies fall faster than lighter ones in the same medium, whereas Galileo claimed that in a vacuum all bodies fall with the same speed.

Aristotle and Galileo also did not agree about the nature of energy. Historians of science attribute the first use of the word *energy* to Aristotle (in classical Greek, *energeia*). Historians debate what exactly Aristotle meant by the word *energeia*, with the most common translation being "activity or operation." Historians do agree, however, that Aristotle used the concept of energeia not only to analyze the motion of objects but also to understand ethics and psychology. According to Aristotle, a person's desire to be happy was a type of energeia. Galileo, on the other hand, provided a more modern view of energy and suggested that energy was a property of objects. Thus, he decoupled the term *energy* from the study of ethics or psychology.

Since Galileo's time, other physicists have contributed to our understanding of forces, motion, and energy. Isaac Newton, in the 17th and early 18th centuries, provided the mathematical foundation for the study of force and motion. In the 19th century, Thomas Young was the first to define *energy* in the formal sense, and Gaspard-Gustave de Coriolis expanded on Young's definition of energy by introducing the concept of kinetic energy. Also in the 19th century, William Rankine was the first to propose the idea of potential energy, and James Prescott Joule and William Thomson (better known as Lord Kelvin) developed the law of conservation of energy, which states that the total energy in a closed system remains constant. We now use these ideas about energy to help understand and explain a wide range of natural phenomena.

In this investigation, you will have an opportunity to use these important ideas about energy to explore the motion of a pendulum. The motion of a pendulum has been studied for hundreds of years. Galileo, for example, used the motion of a pendulum to help quantify his ideas about falling objects. Other scientists have also used the pendulum to investigate forces and energy. One of Galileo's most informative and important investigations about the behavior of a pendulum, however, was when he placed a nail in the path of a swinging pendulum as shown in Figure L21.1. This simple modification to a basic pendulum allowed Galileo to explore the energy of the pendulum bob as it swings back and forth.

Your Task

Use what you know about energy, cause-and-effect relationships in science, and the importance of identifying and explaining patterns in nature to design and carry out an investigation to determine how the height of a pendulum swing changes after coming into contact with a nail in its path.

The guiding question of this investigation is, *How does placing a nail in the path of a pendulum affect the height of a pendulum swing?*

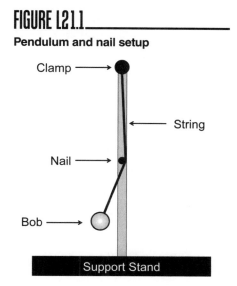

FIGURE L21.1

Pendulum and nail setup

Materials

You may use any of the following materials during your investigation:

- Safety glasses or goggles (required)
- Video camera
- Computer or tablet with video analysis software

- Support stand Clamps Nail
- Hanging mass set
- String
- Meterstick

Safety Precautions

Follow all normal lab safety rules. In addition, take the following safety precautions:

1. Wear sanitized safety glasses or goggles during lab setup, hands-on activity, and takedown.

2. Handle nails with care. Nail ends can be sharp and can puncture or scrape skin.

3. Keep fingers and toes out of the way of moving objects.

4. Wash hands with soap and water after completing the lab.

Investigation Proposal Required?   ☐ Yes   ☐ No

Getting Started

To answer the guiding question, you will need to design and carry out an investigation to determine how the height of a pendulum swing changes after the string comes into contact with a nail as the bob swings back and forth. Be sure to focus only on the height of a pendulum swing during the first few oscillations, because air resistance will cause the pendulum to eventually slow down and come to rest. It is also important to examine pendulums with different lengths of string and different bob masses during your investigation. With these issues in mind, you can now decide what type of data you need to collect, how you will collect it, and how you will analyze it.

To determine *what type of data you need to collect*, think about the following questions:

- What are the boundaries and components of the system you are studying?
- How can you describe the components of the system quantitatively?
- How could you keep track of changes in this system quantitatively?
- What could cause a change in the height of the pendulum swing?
- How will you determine the height of pendulum swing?
- Is it useful to track how energy flows into, out of, or within this system?
- What else will you need to measure during the investigation?

To determine *how you will collect the data*, think about the following questions:

- What type of research design needs to be used to establish a cause-and-effect relationship?
- How could you track the flow of energy within this system?
- How long will you need to observe the pendulum swinging back and forth?
- What types of comparisons will be useful?
- How will you vary the length of the string and the mass of the bob?
- What will be the reference point for your measurements?
- What equipment will you need to use in order to make the measurements you need?
- How will you make sure that your data are of high quality (i.e., how will you reduce error)?
- How will you keep track of and organize the data you collect?

To determine *how you will analyze the data*, think about the following questions:

- What types of patterns might you look for as you analyze your data?
- How could you use mathematics to determine if there is a difference between conditions?
- How precise is your video (frames per second)?
- What type of table or graph could you create to help make sense of your data?

Connections to the Nature of Scientific Knowledge and Scientific Inquiry

As you work through your investigation, you may want to consider

- the difference between observations and inferences in science, and
- how scientific knowledge changes over time.

Initial Argument

Once your group has finished collecting and analyzing your data, your group will need to develop an initial argument. Your argument must include a claim, evidence to support your claim, and a justification of the evidence. The *claim* is your group's answer to the guiding question. The *evidence* is an analysis and interpretation of your data. Finally, the *justification* of the evidence is why your group thinks the evidence matters. The justification of the evidence is important because scientists can use different kinds of evidence to support their claims. Your group will create your initial argument on a whiteboard. Your whiteboard should include all the information shown in Figure L21.2.

FIGURE L21.2 _____
Argument presentation on a whiteboard

The Guiding Question:	
Our Claim:	
Our Evidence:	Our Justification of the Evidence:

Argumentation Session

The argumentation session allows all of the groups to share their arguments. One or two members of each group will stay at the lab station to share that group's argument, while the other members of the group go to the other lab stations to listen to and critique the other arguments. This is similar to what scientists do when they propose, support, evaluate, and refine new ideas during a poster session at a conference. If you are presenting your group's argument, your goal is to share your ideas and answer questions. You should also keep a record of the critiques and suggestions made by your classmates so you can use this feedback to make your initial argument stronger. You can keep track of specific critiques and suggestions for improvement that your classmates mention in the space below.

Critiques about our initial argument and suggestions for improvement:

LAB 21

If you are critiquing your classmates' arguments, your goal is to look for mistakes in their arguments and offer suggestions for improvement so these mistakes can be fixed. You should look for ways to make your initial argument stronger by looking for things that the other groups did well. You can keep track of interesting ideas that you see and hear during the argumentation in the space below. You can also use this space to keep track of any questions that you will need to discuss with your team.

Interesting ideas from other groups or questions to take back to my group:

Once the argumentation session is complete, you will have a chance to meet with your group and revise your initial argument. Your group might need to gather more data or design a way to test one or more alternative claims as part of this process. Remember, your goal at this stage of the investigation is to develop the best argument possible.

Report

Once you have completed your research, you will need to prepare an *investigation report* that consists of three sections. Each section should provide an answer to the following questions:

1. What question were you trying to answer and why?

2. What did you do to answer your question and why?

3. What is your argument?

Your report should answer these questions in two pages or less. This report must be typed, and any diagrams, figures, or tables should be embedded into the document. Be sure to write in a persuasive style; you are trying to convince others that your claim is acceptable or valid!

Checkout Questions

Lab 21. Conservation of Energy and Pendulums: How Does Placing a Nail in the Path of a Pendulum Affect the Height of a Pendulum Swing?

1. Pictured at right is a pendulum. Let **h** = 0 represent the height when the bob is at equilibrium and acceleration due to gravity is 9.8 m/s^2 With these facts, calculate the kinetic energy of the bob when it is at the equilibrium position, assuming the pendulum is released from the point where **h** = 10 cm.

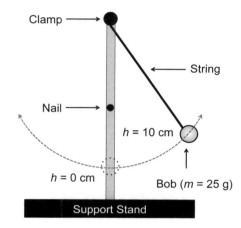

2. The length of the pendulum has an effect on the height of the bob after contacting and sweeping through the nail (assuming the initial height is not higher than the nail).

   a. I agree with this statement.

   b. I disagree with this statement.

   Explain your answer, using the findings from your investigation about placing a nail in the path of a pendulum.

LAB 21

3. There is no difference between observations and inferences in science.

    a. I agree with this statement.

    b. I disagree with this statement.

Explain your answer, using an example from your investigation about placing a nail in the path of a pendulum.

4. Scientific knowledge can change over time.

    a. I agree with this statement.

    b. I disagree with this statement.

Explain your answer, using an example from your investigation about placing a nail in the path of a pendulum.

5. Scientists often look for or attempt to identify patterns in nature. Explain why this is a useful practice, using an example from your investigation about placing a nail in the path of a pendulum.

6. In science, understanding cause-and-effect relationships is an important goal. Sometimes, scientists hypothesize a causal relationship that is not supported by the data. Does this mean there was something wrong with the investigation? Explain your answer, using an example from your investigation about placing a nail in the path of a pendulum.

7. In a pendulum, energy is transferred from potential to kinetic energy. Discuss the relationship between the potential and kinetic energy of the pendulum as it swings back and forth. Is energy conserved? Why is it important to keep track of the energy as it is transferred from potential to kinetic energy?

Lab Handout

Lab 22. Conservation of Energy and Wind Turbines: How Can We Maximize the Amount of Electrical Energy That Will Be Generated by a Wind Turbine Based on the Design of Its Blades?

Introduction

The United States relies heavily on fossil fuels, such as oil, coal, and natural gas, for energy. Americans burn fossil fuels to power their cars, to heat and light their homes, and to run their consumer electronics and household appliances. Unfortunately, when the supply of fossil fuels declines, it can lead to an energy crisis. In 1973, for example, the Organization of the Petroleum Exporting Companies (OPEC) declared an oil embargo and drastically reduced the amount of oil they sent to the United States. This six-month embargo caused widespread gasoline shortages and significant price increases. A second oil crisis in the United States occurred in 1979 as a result of a decrease in the output of oil from the Middle East along with price increases. This decrease in output resulted in gasoline shortages and long lines at gasoline stations throughout the United States (see Figure L22.1). These events led to a political push for policies that would make America more energy independent. A country that is energy independent does not need to import oil or other forms of energy from other countries to meet the energy demands of its citizens.

A long line at a gas station in Maryland as a result of the 1979 oil crisis

More recently, evidence that the burning of fossil fuels contributes to climate change has exacerbated the need to develop new energy sources. To combat the effects of climate change, scientists have been working on ways to harness and use renewable energy sources. Renewable energy sources are forms of energy that are continuously replenished. Solar, tidal, and geothermal energy are examples of renewable energy sources. Furthermore, all renewable energy sources are clean, which means they do not contribute to climate change. These sources of energy, however, are often difficult to access without the development of new technologies.

One of the more promising renewable energy sources is wind. Wind is just the movement of a large number of air particles from one area to another. In the United States,

Conservation of Energy and Wind Turbines

How Can We Maximize the Amount of Electrical Energy That Will Be Generated by a Wind Turbine Based on the Design of Its Blades?

wind energy is a very attractive option because the Great Plains states of Texas, Oklahoma, Kansas, Nebraska, and the Dakotas have vast open spaces where wind blows for extended periods of time. Wind energy takes advantage of the law of conservation of energy to convert the kinetic energy of air particles (gases such nitrogen [N_2], oxygen [O_2], water vapor, and others) into electrical energy. To achieve this conversion, several huge wind turbines are built to create a wind farm (see Figure L22.2). When air particles collide with the blades on the turbine, they transfer some momentum to the blades. This momentum transfer causes the turbine to spin, resulting in an increase in the kinetic energy of the turbine blades. The turbine then turns a generator, producing an electric current from a transfer of kinetic energy to electric energy. Wires carry the current to other locations, for use in homes and businesses.

FIGURE L22.2

A wind farm

When designing a wind turbine, scientists and engineers must account for a variety of factors related to how and where wind blows. As mentioned earlier, wind farms are best placed in the Great Plains, yet only a small percentage of people live in this geographic location, so scientists and engineers must also think about how to move the energy generated to more populous areas. Another factor is the height of the turbine. Wind tends to blow faster at higher altitudes, meaning the height of the turbine affects the amount of energy produced. Other potential factors that may influence the amount of electrical energy produced include the angle of the blades relative to the wind, the number of blades on the turbine, and the shape and mass of the turbine blades.

Your Task

Use what you know about the conservation of energy, systems, models, and structure and function to test the effect of changes to the design of wind turbine blades on the amount of electrical energy produced by the wind turbine. Your goal is to determine how the angle, number, and shape of the blades that are attached to the wind turbine affect the amount of electrical energy that the wind turbine is able to produce.

The guiding question of this investigation is, *How can we maximize the amount of electrical energy that will be generated by a wind turbine based on the design of its blades?*

LAB 22

Materials

You may use any of the following materials during your investigation:

- Safety glasses or goggles (required)
- Wind turbine kit with adjustable blades
- Fan to generate wind
- Multimeter or galvanometer
- Electric wires
- Lightbulbs
- Ruler
- Protractor

Safety Precautions

Follow all normal lab safety rules. In addition, take the following safety precautions:

1. Wear sanitized safety glasses or goggles during lab setup, hands-on activity, and takedown.

2. Handle blades with care. They can puncture skin.

3. Electric wire may get hot. Use caution when touching exposed parts of wires.

4. Handle lightbulbs with care. They can get hot and burn skin; also, they are fragile and can break easily, causing a sharp hazard that can cut or puncture skin.

5. Keep fingers and toes out of the way of moving objects.

6. Wash hands with soap and water after completing the lab.

Investigation Proposal Required?   ☐ Yes   ☐ No

Getting Started

You will need to design and carry out at least three different experiments to determine how to maximize the amount of electrical energy that is generated by a wind turbine based on the design of its blades. You will need to conduct three different experiments because you will need to be able to answer the following questions before you can develop an answer to the guiding question for this lab:

1. How does changing the number of blades affect the energy output of the turbine?

2. How does changing the shape of the blades affect the energy output of the turbine?

3. How does changing the angle of the blades affect the energy output of the turbine?

It will be important for you to determine what type of data you need to collect, how you will collect it, and how you will analyze it for each experiment, because each experiment is slightly different.

To determine *what type of data you need to collect,* think about the following questions:

- What are the boundaries and components of the system you are studying?
- How do the components of the system interact with each other?
- How can you describe the components of the system quantitatively?
- How could you keep track of changes in this system quantitatively?
- How might the structure of a wind turbine blade relate to its function?
- How might changes to the structure of a wind turbine blade affect how it functions?
- Is it useful to track how energy flows into, out of, or within this system?
- What will be the independent variable and the dependent variable for each experiment?

To determine *how you will collect the data,* think about the following questions:

- How will you vary the independent variable during each experiment?
- What will you do to hold the other variables constant during each experiment?
- When will you need to take measurements or observations during each experiment?
- What scale or scales should you use when you take your measurements?

To determine *how you will analyze the data,* think about the following questions:

- What types of calculations will you need to make?
- What types of comparisons will you need to make?
- How could you use mathematics to determine if there is a difference between the groups?
- What type of table or graph could you create to help make sense of your data?

Connections to the Nature of Scientific Knowledge and Scientific Inquiry

As you work through your investigation, you may want to consider

- how scientific knowledge changes over time, and
- how the culture of science, societal needs, and current events influence the work of scientists.

Initial Argument

Once your group has finished collecting and analyzing your data, your group will need to develop an initial argument. Your initial argument needs to include a claim, evidence to support your claim, and a justification of the evidence. The *claim* is your group's answer to the guiding question. The *evidence* is an analysis and interpretation of your data. Finally, the

LAB 22

justification of the evidence is why your group thinks the evidence matters. The justification of the evidence is important because scientists can use different kinds of evidence to support their claims. Your group will create your initial argument on a whiteboard. Your whiteboard should include all the information shown in Figure L22.3.

Argumentation Session

The argumentation session allows all of the groups to share their arguments. One or two members of each group will stay at the lab station to share that group's argument, while the other members of the group go to the other lab stations to listen to and critique the other arguments. This is similar to what scientists do when they propose, support, evaluate, and refine new ideas during a poster session at a conference. If you are presenting your group's argument, your goal is to share your ideas and answer questions. You should also keep a record of the critiques and suggestions made by your classmates so you can use this feedback to make your initial argument stronger. You can keep track of specific critiques and suggestions for improvement that your classmates mention in the space below.

Critiques about our initial argument and suggestions for improvement:

If you are critiquing your classmates' arguments, your goal is to look for mistakes in their arguments and offer suggestions for improvement so these mistakes can be fixed. You should look for ways to make your initial argument stronger by looking for things that the other groups did well. You can keep track of interesting ideas that you see and hear during the argumentation in the space below. You can also use this space to keep track of any questions that you will need to discuss with your team.

FIGURE L22.3

Argument presentation on a whiteboard

The Guiding Question:	
Our Claim:	
Our Evidence:	Our Justification of the Evidence:

Interesting ideas from other groups or questions to take back to my group:

Once the argumentation session is complete, you will have a chance to meet with your group and revise your initial argument. Your group might need to gather more data or design a way to test one or more alternative claims as part of this process. Remember, your goal at this stage of the investigation is to develop the best argument possible.

Report

Once you have completed your research, you will need to prepare an *investigation report* that consists of three sections. Each section should provide an answer to the following questions:

1. What question were you trying to answer and why?

2. What did you do to answer your question and why?

3. What is your argument?

Your report should answer these questions in two pages or less. This report must be typed, and any diagrams, figures, or tables should be embedded into the document. Be sure to write in a persuasive style; you are trying to convince others that your claim is acceptable or valid!

Checkout Questions

Lab 22. Conservation of Energy and Wind Turbines: How Can We Maximize the Amount of Electrical Energy That Will Be Generated by a Wind Turbine Based on the Design of Its Blades?

1. The electrical energy produced by a wind turbine originates as solar energy. Describe the processes that transfer solar energy from the Sun into electrical energy in the wires produced by the turbine.

2. Using concepts of (1) conservation of momentum, (2) conservation of energy, (3) the definition of momentum as a vector quantity, and (4) the definition of energy as a scalar quantity, explain why there is a maximum value for the energy output for one of the variables you tested.

3. People view some research as being more important than other research because of current events or what is important in society.

    a. I agree with this statement.
    b. I disagree with this statement.

    Explain your answer, using an example from your investigation about wind turbines.

4. Scientific knowledge, once it has been proven true, does not change.

    a. I agree with this statement.
    b. I disagree with this statement.

    Explain your answer, using an example from your investigation about wind turbines.

5. How something is structured can affect that object's function.

   a. I agree with this statement.
   b. I disagree with this statement.

   Explain your answer, using an example from your investigation about wind turbines.

6. Scientists often need to track how matter moves into and within a system. Explain why this is important, using an example from your investigation about wind turbines.

7. Scientists often need to define a system under study and then create a model during an investigation. Explain why models of systems are useful in science, using an example from your investigation about wind turbines.

Application Lab

Lab Handout

Lab 23. Power: Which Toy Car Has the Engine With the Greatest Horsepower?

Introduction

With the onset of the Industrial Revolution, many people wanted a way to compare the capabilities of new machines with existing ways of doing things. For example, when James Watt started building and selling steam engines in the late 18th century, one of his customers asked Watt to compare the power of the engine with the power of a horse. The customer, in other words, wanted to know if the new steam engine could do the same things a horse could do. Watt used the term *horsepower* as a convenient way of comparing the power of his steam engine with the power of a horse. Figure L23.1 shows a painting by Carl Rakeman of a race in 1830 between a horse-drawn cart and the first steam locomotive built in the United States, which was called Tom Thumb. This race was an important event leading to the rise of the railroad industry in America because it demonstrated that a steam-powered locomotive could outperform a horse. Since then, *horsepower* has been given a formal definition as a unit of measure, with 1 horsepower (hp) equal to 746 watts (W).

FIGURE L23.1

Painting by Carl Rakeman of the Tom Thumb steam-powered locomotive racing a horse

Note: A full-color, high-resolution version of this image is available on the book's Extras page at *www.nsta.org/adi-physics1*.

Although most countries use the watt or kilowatt as the standard unit of power, in the United States horsepower is still quite common. Most devices that are powered by an engine or motor are advertised based upon their horsepower; these include household devices such as blenders, lawnmowers, and electric garage door openers. The most common use for horsepower is to describe the power of a car engine. Most everyday cars have an engine of approximately 150 hp, but the fastest sports cars can have engines that are up to 600 hp. And race cars can have engines that are over 800 hp.

An independent group must verify the advertised horsepower of a new engine before a company can sell that engine or a vehicle with that engine in it. An independent group can use several different methods to measure the horsepower of an engine that they did not build; these methods are based on the conservation of energy and the relationship between work and power. The two most common methods are (1) to measure how long it takes the engine to lift or pull a mass a given distance and (2) to measure the amount of time it takes the car to reach its maximum velocity when starting from rest and then use the work-energy theorem to determine the maximum horsepower of the engine. In this investigation, you will have an opportunity to develop your own method for measuring the horsepower of a toy car's engine using the work-energy theorem.

Your Task

Use what you know about the conservation of energy, the relationship between work and power, systems and system models, and the importance of tracking the flow of energy in a system to develop a method for measuring the horsepower of a toy car engine. You will then test the engines of several different toy cars using your method to determine which one has the greatest horsepower.

The guiding question of this investigation is, *Which toy car has the engine with the greatest horsepower?*

Materials

You may use any of the following materials during your investigation:

Consumables	Equipment	
• Tape	• Safety glasses or goggles (required)	• Stopwatches
• String or fishing line	• 4 Different types or brands of remote control cars	• Ruler
	• Hanging mass set	• Metersticks
	• Electronic or triple beam balance	

If you have access to the following equipment, you may also consider using a video camera and a computer or tablet with video analysis software.

Safety Precautions

Follow all normal lab safety rules. In addition, take the following safety precautions:

1. Wear sanitized safety glasses or goggles during lab setup, hands-on-activity, and takedown.

2. Keep fingers and toes out of the way of moving objects.

3. Wash hands with soap and water after completing the lab.

LAB 23

Getting Started

To answer the guiding question, you will need to create a method that you can use to determine the horsepower of the engine found inside a toy car. This means you will need to determine the maximum power that the engine is able to produce. *Power* is defined as the rate at which work is done on an object. The equation for power is $P = W/t$, where W is work and t is time. The term *work* is used to describe any situation when a force acts on an object over a displacement of that object. For a force to qualify as having done work on an object, there must be a displacement and the force must be either parallel to the displacement (e.g., a force to the right with a displacement to the right) or antiparallel to the displacement (e.g., a force to the right with a displacement to the left). The equation for work is $W = \mathbf{F}d$, where $\mathbf{F}$ is force and $\mathbf{d}$ is displacement. Finally, and perhaps most important in terms of your goal for this investigation, energy cannot be created or destroyed—it just changes form as objects move or as it transfers from one object to another one. You will need to use these fundamental ideas as the foundation for the method you will develop, but there may also be other ideas that you need to use.

To determine *what type of data you need to collect*, think about the following questions:

- What are the boundaries and components of the system under study?
- How can you describe the components of the system quantitatively?
- How could you keep track of changes in this system quantitatively?
- Is it useful to track how energy flows into, out of, or within this system?
- How might the structure of the toy car affect its function?
- How might the structure of the test of horsepower affect its function?

To determine *how you will collect the data*, think about the following questions:

- What types of equipment will you need to use and how will you use it?
- How will you track the flow of energy into, out of, and within the system under study?
- What scale or scales should you use when you take your measurements?
- How will you make sure that your data are of high quality (i.e., how will you reduce error)?
- How will you keep track of the data you collect?

To determine *how you will analyze the data*, think about the following questions:

- What type of calculations will you need to make?
- What types of comparisons will you need to make?

- What type of table or graph could you create to help make sense of your data?

Connections to the Nature of Scientific Knowledge and Scientific Inquiry

As you work through your investigation, you may want to consider

- the difference between observations and inferences in science, and
- the role of imagination and creativity in science.

Initial Argument

Once your group has finished collecting and analyzing your data, your group will need to develop an initial argument. Your argument must include a claim, evidence to support your claim, and a justification of the evidence. The *claim* is your group's answer to the guiding question. The *evidence* is an analysis and interpretation of your data. Finally, the *justification* of the evidence is why your group thinks the evidence matters. The justification of the evidence is important because scientists can use different kinds of evidence to support their claims. Your group will create your initial argument on a whiteboard. Your whiteboard should include all the information shown in Figure L23.2.

FIGURE L23.2

Argument presentation on a whiteboard

The Guiding Question:	
Our Claim:	
Our Evidence:	Our Justification of the Evidence:

Argumentation Session

The argumentation session allows all of the groups to share their arguments. One or two members of each group will stay at the lab station to share that group's argument, while the other members of the group go to the other lab stations to listen to and critique the other arguments. This is similar to what scientists do when they propose, support, evaluate, and refine new ideas during a poster session at a conference. If you are presenting your group's argument, your goal is to share your ideas and answer questions. You should also keep a record of the critiques and suggestions made by your classmates so you can use this feedback to make your initial argument stronger. You can keep track of specific critiques and suggestions for improvement that your classmates mention in the space below.

Critiques about our initial argument and suggestions for improvement:

If you are critiquing your classmates' arguments, your goal is to look for mistakes in their arguments and offer suggestions for improvement so these mistakes can be fixed. You should look for ways to make your initial argument stronger by looking for things that the other groups did well. You can keep track of interesting ideas that you see and hear during the argumentation in the space below. You can also use this space to keep track of any questions that you will need to discuss with your team.

Interesting ideas from other groups or questions to take back to my group:

Once the argumentation session is complete, you will have a chance to meet with your group and revise your initial argument. Your group might need to gather more data or design a way to test one or more alternative claims as part of this process. Remember, your goal at this stage of the investigation is to develop the best argument possible.

Report

Once you have completed your research, you will need to prepare an *investigation report* that consists of three sections. Each section should provide an answer to the following questions:

1. What question were you trying to answer and why?

2. What did you do to answer your question and why?

3. What is your argument?

Your report should answer these questions in two pages or less. This report must be typed, and any diagrams, figures, or tables should be embedded into the document. Be sure to write in a persuasive style; you are trying to convince others that your claim is acceptable or valid!

Checkout Questions

Lab 23. Power: Which Toy Car Has the Engine With the Greatest Horsepower?

1. How do the law of conservation of energy and the work-energy theorem help engineers determine the horsepower of a motor?

2. In the figure at right, a person uses a pulley to lift a box off the ground. Assuming that the person lifts the box of mass m with a constant velocity $\mathbf{v}$ and lifts the box a height of $\Delta \mathbf{h}$ in t seconds, create a mathematical expression for the rate at which the person does work on the box.

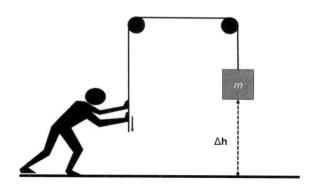

3. Scientists need to be creative and have a good imagination.

    a. I agree with this statement.
    b. I disagree with this statement.

Explain your answer, using an example from your investigation about the horse-power of a toy car.

4. A scientist must first make an observation before he or she can make an inference.

    a. I agree with this statement.
    b. I disagree with this statement.

Explain your answer, using an example from your investigation about the horse-power of a toy car.

5. Scientists often need to define a system under study during an investigation. Explain why it is useful to define a system under study during an investigation, using an example from your investigation about the horsepower of a toy car.

6. Scientists often need to track how matter moves into and within a system. Explain why this is important, using an example from your investigation about the horsepower of a toy car.

IMAGE CREDITS

All images in this book are stock photographs or courtesy of the authors unless otherwise noted below.

Lab 4

Figure L4.1: Martin23230, Wikimedia Commons, CC BY-SA 3.0. *https://commons.wikimedia.org/wiki/File:Americas_(orthographic_projection).svg*

Map of North America: Wikimedia Commons, GFDL 1.2. *https://commons.wikimedia.org/wiki/File:North_America_(orthographic_projection).svg*

Lab 6

Figure L6.1: U.S. Patent and Trademark Office, Public domain. *https://tinyurl.com/zvwyrcm*

Lab 8

Figure L8.1: Mj-bird, Wikimedia Commons, CC BY-SA 3.0. *https://commons.wikimedia.org/wiki/File:Internal_combustion_engine_pistons_of_partial_cross-sectional_view.jpg*

Lab 9

Figure L9.1: U.S. Air Force, Flickr, *https://c1.staticflickr.com/5/4068/4317655660_61a60f6576_b.jpg*

Lab 11

Figure L11.1: FaceMePLS, Wikimedia Commons, CC BY-SA 2.0. *https://commons.wikimedia.org/wiki/File:Kermis_Malieveld_Den_Haag.jpg*

Lab 13

Figure L13.2: Chetvorno, Wikimedia Commons, Public domain. *https://commons.wikimedia.org/wiki/Category:Diagrams_of_pendulums#/media/File:Simple_gravity_pendulum.svg*

Figure L13.1: Sjoerd22, Wikimedia Commons, Public domain. *https://commons.wikimedia.org/wiki/File:H6_clock.jpg*

Lab 14

Figure L14.1: Svjo, Wikimedia Commons, CC BY-SA 3.0. *https://commons.wikimedia.org/wiki/File:Mass-spring-system.png*

Lab 16

Figure L16.1: Max Andrews, Wikimedia Commons, CC BY-SA 3.0. *https://upload.wikimedia.org/wikipedia/commons/6/67/Concussion_Anatomy.png*

Figure L16.2: Damon J. Moritz, Public domain. *https://commons.wikimedia.org/wiki/File:US_Navy_031230-N-9693M-004_Navy_linebacker_Bobby_McClarin_sacks_Texas_Tech_quarterback_B.J._Symons_during_the_EV1.Net_Houston_Bowl_at_Reliant_Stadium_in_Houston,_Texas.jpg*

Lab 22

Figure L22.1: Warren K. Leffler, Wikimedia Commons, Public domain. *https://en.wikipedia.org/wiki/1979_energy_crisis#/media/File:Line_at_a_gas_station,_June_15,_1979.jpg*

Lab 23

Figure L23.1: Federal Highway Administration, U.S. Department of Transportation, Public domain. *www.fhwa.dot.gov/rakeman/1830.htm*